CHINA UAV, UCAV, SUICIDE DRONES, & SPACEPLANES

武装无人机 (Armed Drones)

Alexandre Zanfirov

Revised 18 October 2020

DISCLAIMERS

The information and opinions contained in this document are provided "as is" and without any warranties or guarantees. Reference herein to any specific commercial products, process, or service by trade name, trademark, manufacturer, or otherwise does not constitute or imply its endorsement, recommendation, or favoring by the United States Government, and this guidance shall not be used for advertising or product endorsement purposes.

The statements of fact, opinion, or analysis expressed in this manuscript are those of the author and do not reflect the official policy or position of the Defense Intelligence Agency, the Department of Defense, or the U.S. Government. Review of the material does not imply DIA, DoD, or the U.S. Government endorsement of factual accuracy or opinion.

Copyright © 2020 - 4th Watch Publishing Co.

Other books we publish on Amazon.com

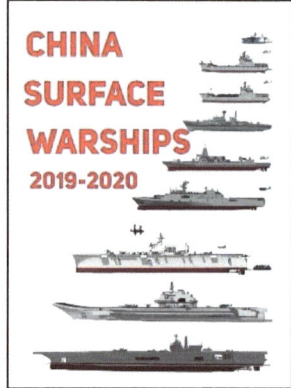

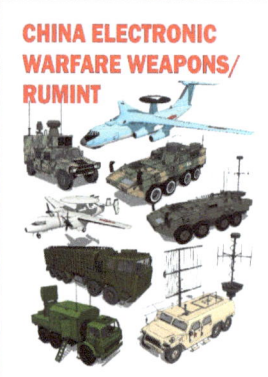

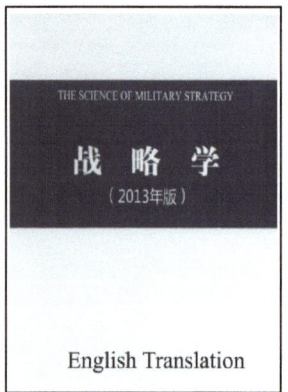
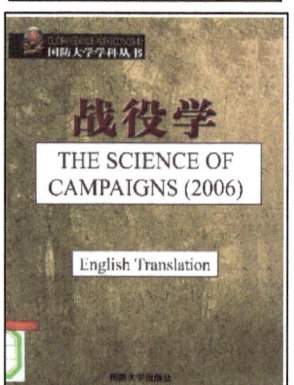

FOREWARD

The PLA has fielded a range of UAVs across all four services, the PLA Army (PLAA), Navy (PLAN), Air Force (PLAAF), and Rocket Force (PLARF/former Second Artillery Force); as well as the Strategic Support Force (PLASSF).[1] A review of open sources revealed over 1,600 different UAVs of all type. Can't do them all and unfortunately, some programs are so secret, there are no photos available. I've managed to cull this list to only those vehicles I believe are of some military significance, either because they can carry a substantial payload or are technologically advanced (such as mature artificial intelligence or autonomous swarm capability). That rules out blimps, *most* helicopters, and *most* Micro Aerial Vehicles (MAV). I tried to focus on UCAVs specifically designed for reconnaissance and strike (察打一体).

NOTE: I decided to include China's next generation spaceplanes because although military uses are banned under the **Outer Space Treaty of 1967** (China Accession, London, 12 January 1984.), spaceplanes have potential military applications. Activities forbidden under the treaty include:
- Placing in orbit around the Earth or other celestial bodies any nuclear weapons or objects carrying WMD.
- Installing WMD on celestial bodies or stationing WMD in outer space in any other manner.

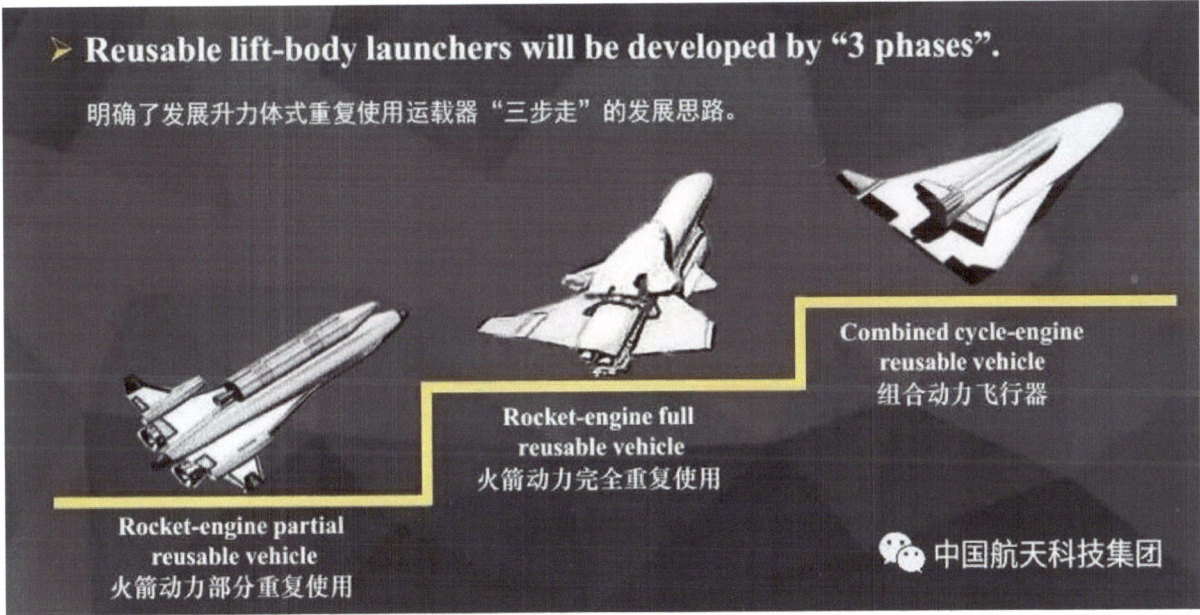

RUMINT is shown in red.

---

[1] Kania, Elsa, *"The PLA's Unmanned Aerial Systems New Capabilities for a "New Era" of Chinese Military Power"*, China Aerospace Studies Institute, Air University.

ACRONYMS

CASC	China Aerospace Science and Technology Corporation
COMINT	communications intelligence
ELINT	electronic intelligence
EO	electro-optical
EXP	experimental
FMS	foreign military sale
GCS	ground control station
HALE	high altitude, long endurance (altitude 18 km, jet engine)
HTHL	horizontal takeoff, horizontal landing
IAI	Israel Aerospace Industries
ISR	intelligence, surveillance, and reconnaissance
LALE	low-altitude, long endurance (altitude up to 3,000 m)
LOS	line of sight
MALE	medium altitude, long endurance (altitude <9,000 ft, 24 hr flight, propeller)
MAME	medium altitude medium endurance
MASINT	measurement and signature intelligence
MAV	micro air vehicle (typically unarmed)
MUMT	Manned-Unmanned Teaming
MTCR	Missile Technology Control Regime
NAS	National Airspace System
OPA	optionally piloted aircraft
RATO	rocket-assisted take off
RPA	remotely piloted aircraft
RUMINT	rumor intelligence
SAR	synthetic aperture radar
SATCOM	satellite communications
SEAD	suppression of enemy air defenses
SIGINT	signal intelligence
SSTO	single-stage to orbit
TBCC	turbine-based combined cycle
TSTO	two-stage to orbit
UAS	unmanned aircraft system
UAV	unmanned aerial vehicle
UCAV	unmanned combat aerial vehicle
VLRAAM	very-long-range air-to-air missile
VTHL	vertical takeoff, horizontal landing
VTVL	vertical take-off, vertical landing
WMD	weapon of mass destruction
XTER	Turbine Ejector-Ramjet Combined Cycle

Table of Contents

UAV SWARM LAUNCH VEHICLE 0
Shenlong Reusable Robotic VTHL Spaceplane ("Divine Dragon") 1
Tengyun Reusable Robotic TSTO, HTHL ("Cloud Climber") 3
Aotian-1 SECRET Spaceplane Concept 3
iSpace Suborbital TSTO, VTHL Spaceplane 3
Qinlong Suborbital TSTO, VTHL Spaceplane 3
Tianxing-1 Suborbital SSTO, VTHL Spaceplane 4
Tianxing-2 Suborbital VTHL Spaceplane 4
Tianxing-3 Suborbital TSTO, VTHL Spaceplane 4
Tianxing-4 Suborbital Spaceplane 4
CASC CH-7 "Rainbow-7" Stealth UCAV (X-47B Clone) 5
Aisheng ASN-9 UAV 7
Aisheng ASN-15 Reconnaissance UAV 7
Aisheng ASN-104 Reconnaissance UAV 8
Aisheng ASN-105 Reconnaissance UAV 8
Aisheng ASN-106 High Speed Target UAV 8
Aisheng ASN-206/207 Reconnaissance UAV 8
Aisheng ASN-209 "Silver Eagle" Multi-Purpose UCAV 9
Aisheng ASN-211 UAV 9
Aisheng ASN-212 Border Patrol UAV 9
Aisheng ASN-213 MAV 9
Aisheng ASN-216 Vertical Take-off and Landing UAV 10
Aisheng ASN-217 Electric Hand-Thrown Drone 10
ASN-218 UAV 11
ASN-219/219A "Magpie III" Long- Endurance Reconnaissance UAV 11
Aisheng ASN-229 Reconnaissance/Strike UCAV 11
ASN-301 Anti-Radiation Radar Loitering Munition Suicide Drone 12
Aisheng BZK-600 UCAV 12
Aisheng DCK-006 Reconnaissance UAV 12
Aisheng JWP02 Reconnaissance UAV 12
ASN C-31 VTOL UAV 12
Aisheng WZ-8 High-Speed, High-Altitude Reconnaissance Drone 13
AT200 Cargo Drone 14

AVIC Yun Ying ("Cloud Shadow") HALE Reconnaissance/Strike 14
AVIC 601-S "Dark Sword" Air-to-Air UCAV 15
AVIC 601-S "Sky Crossbow" UCAV 16
AVIC 601-S "Warrior Eagle" 16
AVIC 601-S "Wind Blade" Drone 16
AVIC AW-4 "Shark II" 16
AVIC L-15 "Blue Fox" Target Drone 16
AVIC Short-Tailed Falcon Drone 17
AVIC "Sky Eye" Disposable Artillery Drone 17
AVIC TL-8 "Sky Dragon" Target Drone 17
AVIC "Whirlwind Scout" MAV 18
AVIC XLB "Patroller" UAV 18
AVIC YY-1 "Swift" MAV 18
BESTUAV SY-5 "Divine Eagle" 5 HALE UCAV 19
BIT "Falcon" Experimental Thrust-Vectoring UAV 19
BIT Gun-Launched UAV 19
Blowfish A2 Helicopter Drone 19
BMP LHK 20
BMP YZ-8 UAV 20
BUAA FH-1 20
BUAA Logistics Unmanned Cargo Aircraft 21
BZK-005 "Giant Eagle" MALE/HALE UCAV 21
CADI "Nimble Loong" MALE Short-Range UAV 23
CAIG "Sky Wing" I UAV 23
CAIG "Sky Wing III" HALE UAV 23
CAIG GJ-I UCAV 23
CAIG Wing Loong I "Pterodactyl I" MALE UCAV 24
CAIG Wing Loong II "Pterodactyl II" MALE UCAV 24
CASC CH-3 Fixed Wing UCAV 25
CASC CH-3A Reconnaissance/ Strike UCAV 25
CASC CH-4 Reconnaissance/Strike UCAV 25
CASC CH-5 "Rainbow 5" UCAV 25
CASC CH-91 Fixed-Wing Reconnaissance and Surveillance UAV 26

CASC CH-92 Fixed Wing Reconnaissance and Surveillance UAV 26
CASC CH-802 Small Hand-Thrown Launch Reconnaissance and Surveillance UAV 26
CASC CH-803 Fixed-Wing Reconnaissance and Surveillance UAV 27
CASC CH-805 Training UAV 27
CH-806 Small Long-Endurance Reconnaissance and Surveillance UAV 27
CASC CH-901/BG-201 Kamikaze Drone UCAV 28
CASC "Peace Map" UAV 28
CASC PW-2 Medium and Short-Range UAV 28
CASC SN-3A 28
CASC CH-T4 Solar Drone 29
CASIC HW-600 Reconnaissance/ Strike UCAV 29
CASIC HW-610 Reconnaissance/ Strike UCAV 30
CASIC HW-800 30
CASIC HW-X100 "Soar Cloud" 30
CASIC HW-X200 30
CASIC SF-460 Blade 30
CASIC "Sky Hawk 1" UCAV 30
CASIC "Spiderman" Fighting Drone 31
CASIC TX-1 UAV 31
CASIC WJ-500 UCAV 31
CASIC WJ-600 A/D UCAV 31
CASIC WJ-700 UCAV 31
CAUC XC-1 "Flying Shuttle" Submarine Launched UAV 32
CDADI Coaxial Shape-Varying Rotary Wing Artillery-Launched UAV 32
CDADI "Mad Warrior" Shape-Changing Wing UAV 32
CDADI "Sea Patroller" Ocean Surveillance UAV 32
CDADI VD200 UAV 32
CH-500 Dual-Rotor Helicopter Drone 32
CH-805 Stealth Drone 33
GAIC BZK-007 UCAV 33
GAIC "Guizhou Central" UCAV 33
GAIC "Harrier Hawk II Air Sniper" UCAV 34
GAIC "Harrier Hawk III" UAV 34

GAIC "Sunshine" Reconnaissance UAV 34

GJ-1 MALE UCAV 35

GJ-2 MALE UCAV 35

WZ-2000 Reconnaissance/Strike UCAV 35

HAIG "Blue Fox" UAV 36

HAIG "Nighthawk" UCAV 36

Hongdu GJ-11 "Sharp Sword" 37

Keyuan AD200 "Blue Eagle" UAV 38

LGAA LN60F "Thunderbird" UAV 38

Linkall HK Drone 38

Nanjing CK-20 High-Speed Target Drone 38

NAV Jet-powered Stealth UAV 39

NAV WF170 Tomahawk Simulation Drone 39

NJUAV LY-Z270 39

NUAA "Cloud Leopard" 39

NUAA FX500 "Sky Saker" Rapid Artillery-Directing UAV 39

NUAA FY-E Reconnaissance UAV 40

NUAA "Long Arrow" UAV 40

NUAA RKL 165 UAV 40

NUAA WZ-1 Soar Bird UAV 40

NUST "Sharp Sword" Carrier-based UAV 41

Oxai B4 UAV 41

SG-1 "Star Glory" UCAV 42

SG-1 "Star Shadow" UCAV 42

SULA89 Kamikaze Drone 43

SYAC "Divine Eagle" Counter Stealth HALE UCAV 43

SYAC "Spider-Man" ZZX Hunter Drone 44

SYAC XLB "Patroller" Fixed-Wing UCAV 44

Tengoen TB001 "Twin-Tailed Scorpion" UCAV 44

V750 UCAV 45

UVS U650-A2 Large Amphibious UAV 45

WBZY BW-I Target Drone 45

Winhye UCAV 46

WZ-7 "Soar Dragon" HALE Reconnaissance UCAV 46
WZ-2000 Reconnaissance/Strike UCAV 47
Xian CCKW LJ-I Stealthy Target Drone.. 48
Xinying "Clairvoyance V" UAV 48
XYAST XYB-180 Reconnaissance UAV 48
XYAST KGXY-180 49
XY Aviation UR-J1-001 "Blue Arrow" BA-270.. 49
ZHZ TD220 Coaxial Unmanned Helicopter... 50
ZJ-100 UAV.. 50

UAV SWARM LAUNCH VEHICLE

The China Academy of Electronics and Information Technology (CAEIT) has demonstrated an armored vehicle capable of launching a swarm of 48 suicide drones.

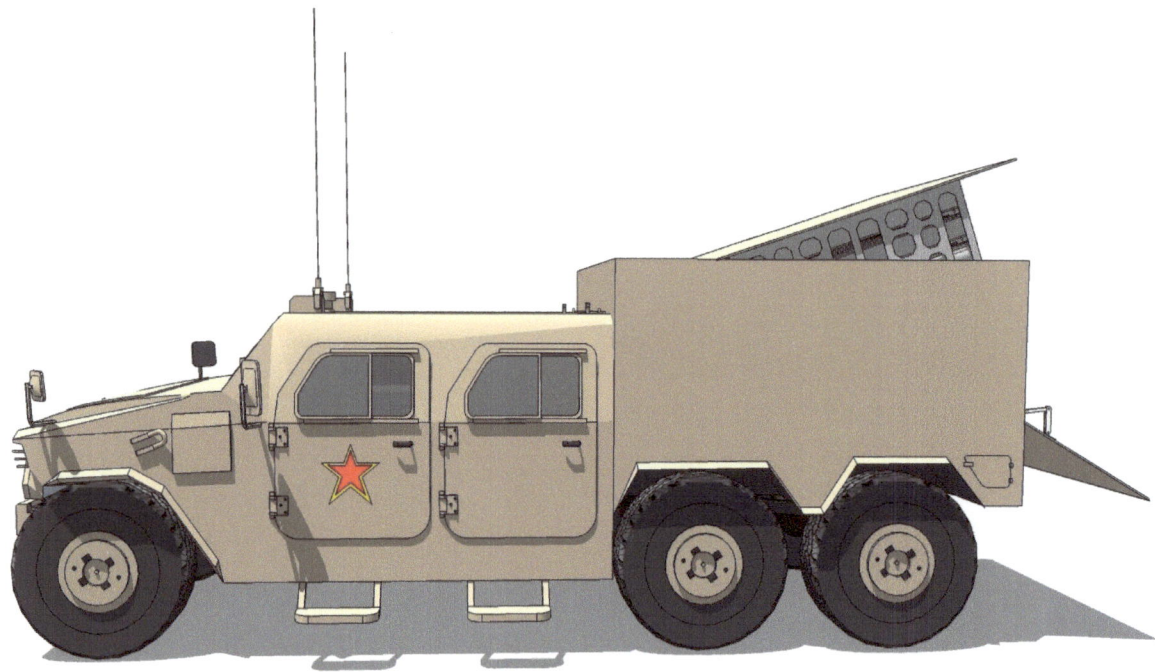

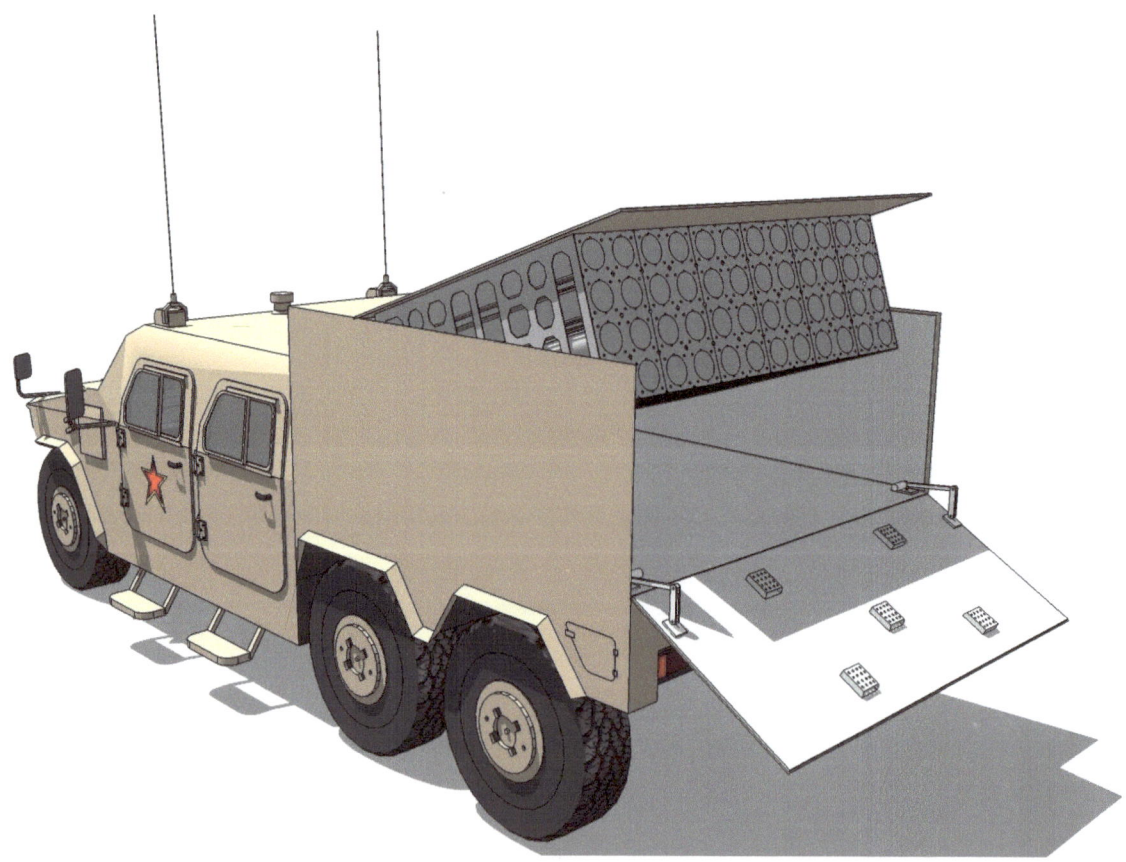

Shenlong Reusable Robotic VTHL Spaceplane ("Divine Dragon")

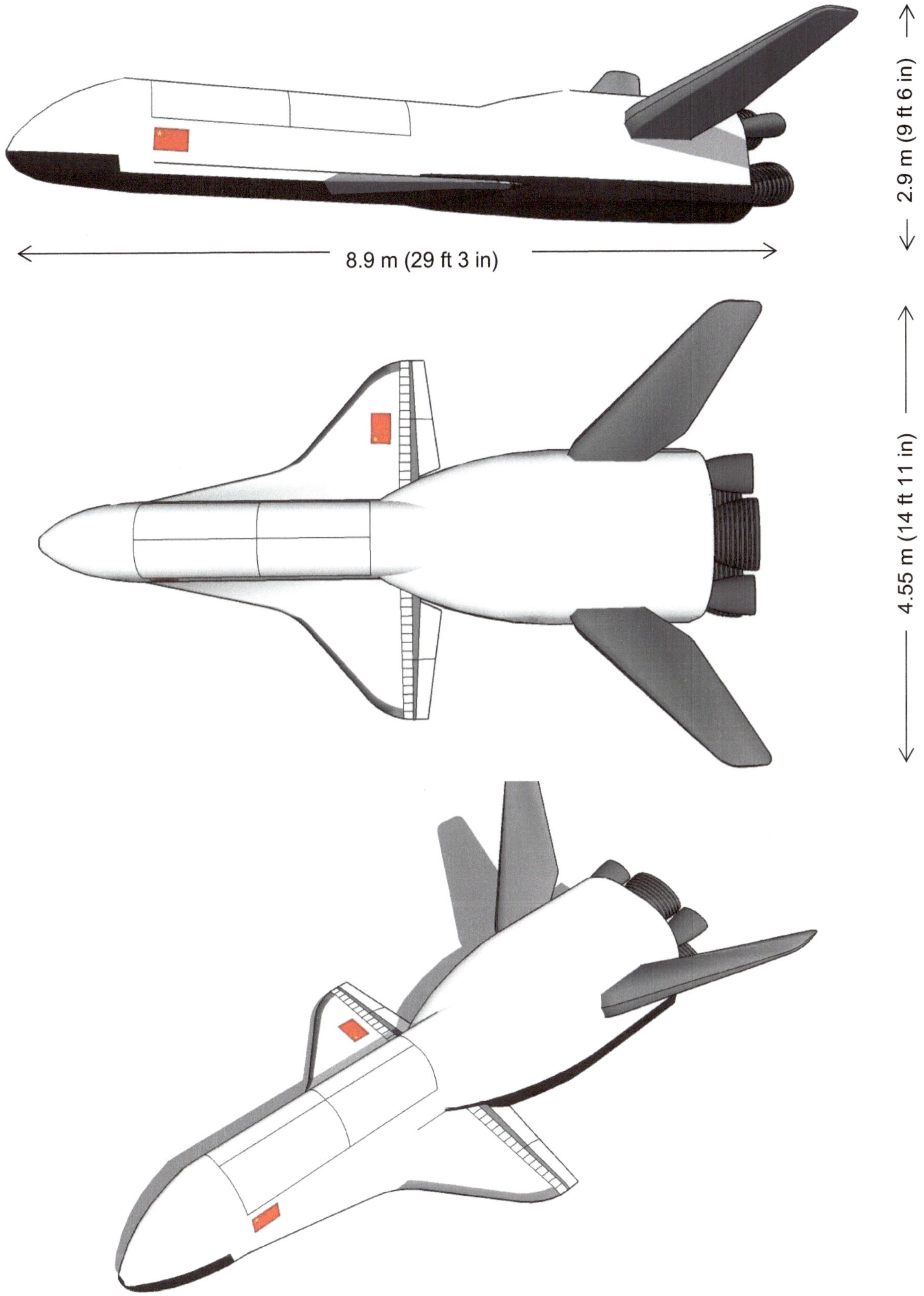

8.9 m (29 ft 3 in)

2.9 m (9 ft 6 in)

4.55 m (14 ft 11 in)

Shenlong Reusable Robotic VTHL Spaceplane ("Divine Dragon") –

Chengdu Shenlong (simplified Chinese: (神龙); traditional Chinese: 神龍; pinyin: shén lóng; lit.: "divine dragon") is a Chinese reusable robotic VTHL (vertical takeoff, horizontal landing) spaceplane designed by AVIC Institute 611 (Chengdu) for long-duration flights in Earth orbit currently in development. It is reportedly a copy of the X-37 operated by the United States. It has been launched on a Long March-2F rocket from the Jiuquan satellite center in Inner Mongolia. It can maneuver at hypersonic speed as needed and can change orbit. It then re-enters Earth's atmosphere and lands as a spaceplane. Because very little information is available on this spacecraft, the drawings and dimensions here are based on the U.S. X-37 Orbital Test Vehicle. An official memo circulating on social media warned staff and visitors to the launch site not to film the lift-off or discuss it online.

Possible missions:

Although claimed to "provide more convenient and cheaper transport for the peaceful use of space in the future", Shenlong could carry any number of payloads, possibly including tungsten "kinetic space weapons" ("Rods from the God")[1], or hypersonic glide missiles. For example, a hypersonic payload(s) using a *dual waverider* design incorporating the Xiamen Turbine Ejector-Ramjet Combined Cycle (XTER) propulsion system and a turbine-based combined cycle (TBCC) engine may be feasible. A *dual waverider* hypersonic payload could reach speeds of Mach 4 to 6 with varying angles of attack.

---

[1] "Rods from the God" is based on the concept of creating man-made meteorites that can be guided towards the enemy. Reaching speeds of 7,000 mph, they would hit the ground with the force of a small nuclear weapon.

Tengyun Reusable Robotic TSTO, HTHL ("Cloud Climber") Spaceplane

The China Aerospace Science and Industry Corporation has been working on a second reusable space plane which is reportedly a larger TSTO, HTHL carrier aircraft with winged 1st and 2nd stages, and a smaller vessel (possibly at Mianyang, Sichuan province, which is also known as **Base 29**). 1st stage propulsion reportedly is a TBCC, both stages are reusable.

Aotian-1 SECRET Spaceplane Concept
Very little information available as of yet. Stay tuned.

iSpace Suborbital TSTO, VTHL Spaceplane
iSpace development of two-stage suborbital VTHL vehicle with unknown payload.
Very little information available.

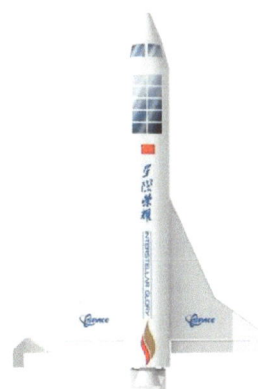

Qinlong Suborbital TSTO, VTHL Spaceplane
Landspace Transportation Co. development of reusable TSTO, VTHL vehicle with (10) taikonauts[2] (?) payload.
Propulsion unknown.
Presented in 2018, status unknown.
Very little information available.

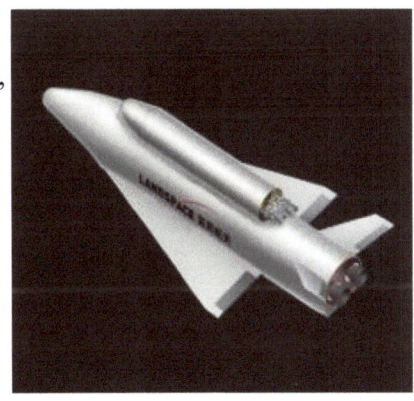

---

[2] A Chinese astronaut, derived from taikong, the Chinese word for space.

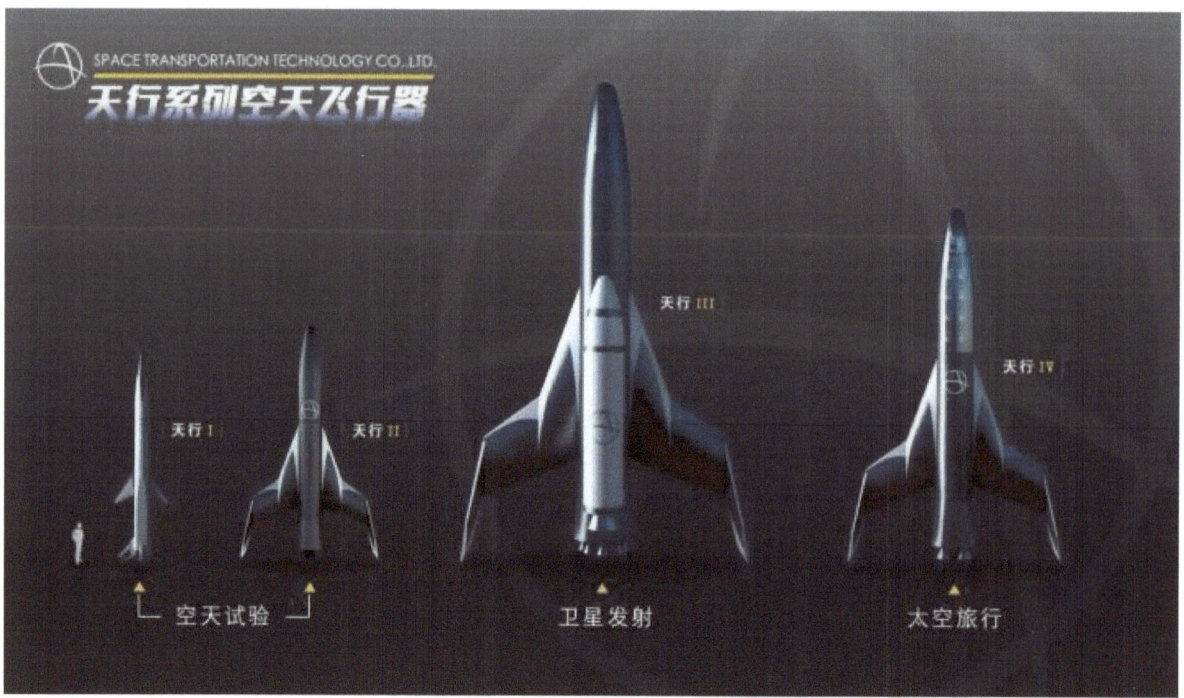

Tianxing-1 Suborbital SSTO, VTHL Spaceplane

Space Transportation Co. development of (天行) reusable single-stage suborbital VTHL vehicle with unknown payload.
Propulsion unknown.
Status unknown.
Very little information available.

Tianxing-2 Suborbital VTHL Spaceplane

Space Transportation Co. development of (天行) reusable suborbital VTHL vehicle with unknown payload.
Propulsion unknown.
Status unknown.
Very little information available.

Tianxing-3 Suborbital TSTO, VTHL Spaceplane

Space Transportation Co. development of (天行) reusable suborbital TSTO VTHL vehicle with unknown payload.
Propulsion unknown.
Status unknown.
Very little information available.

Tianxing-4 Suborbital Spaceplane

A twinkle in someone's eye at this point. No (天行) info available

CASC CH-7 "Rainbow-7" Stealth UCAV (X-47B Clone)

CASC "Rainbow 7" (Cai Hong CH-7) new generation of Chinese bombers to replace the H-6K bomber. It is a flying wing configuration and is smaller than a U.S. B2. Weapons are cruise missiles or bombs. The engine is probably D30KP2.

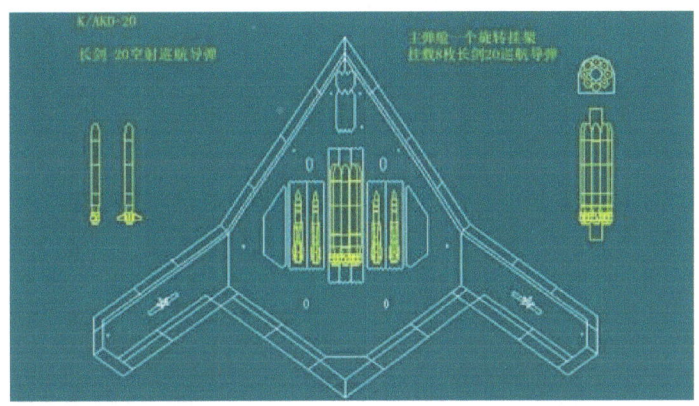

Wingspan - 22 m
Range - 4,000 km
Altitude - 13,000 m
Speed - 800 km/h
Estimated weight - 12 to 13 tons

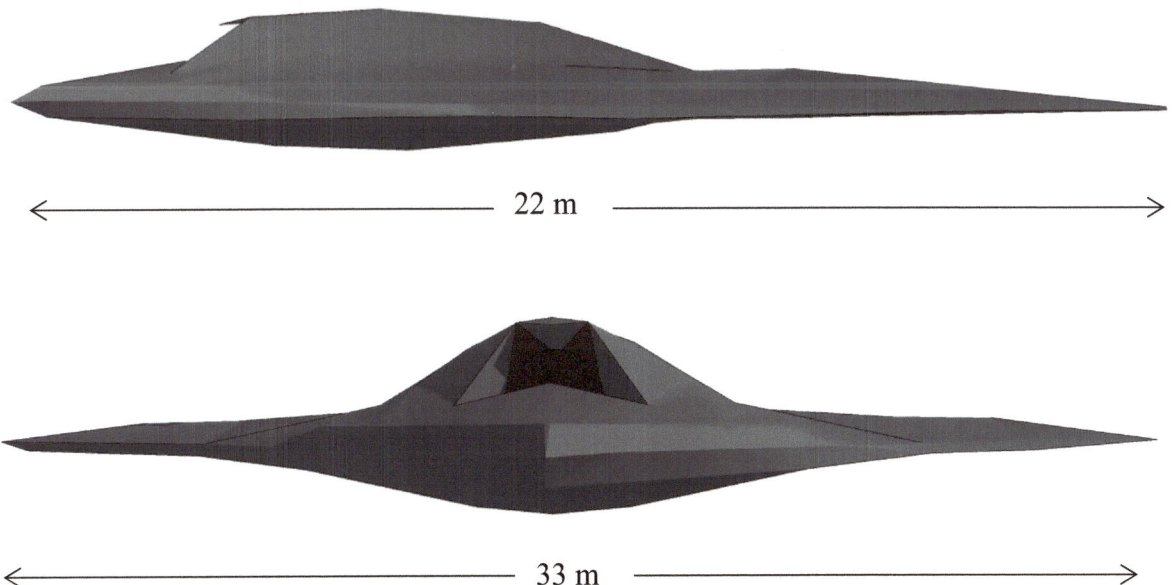

CASC CH-7

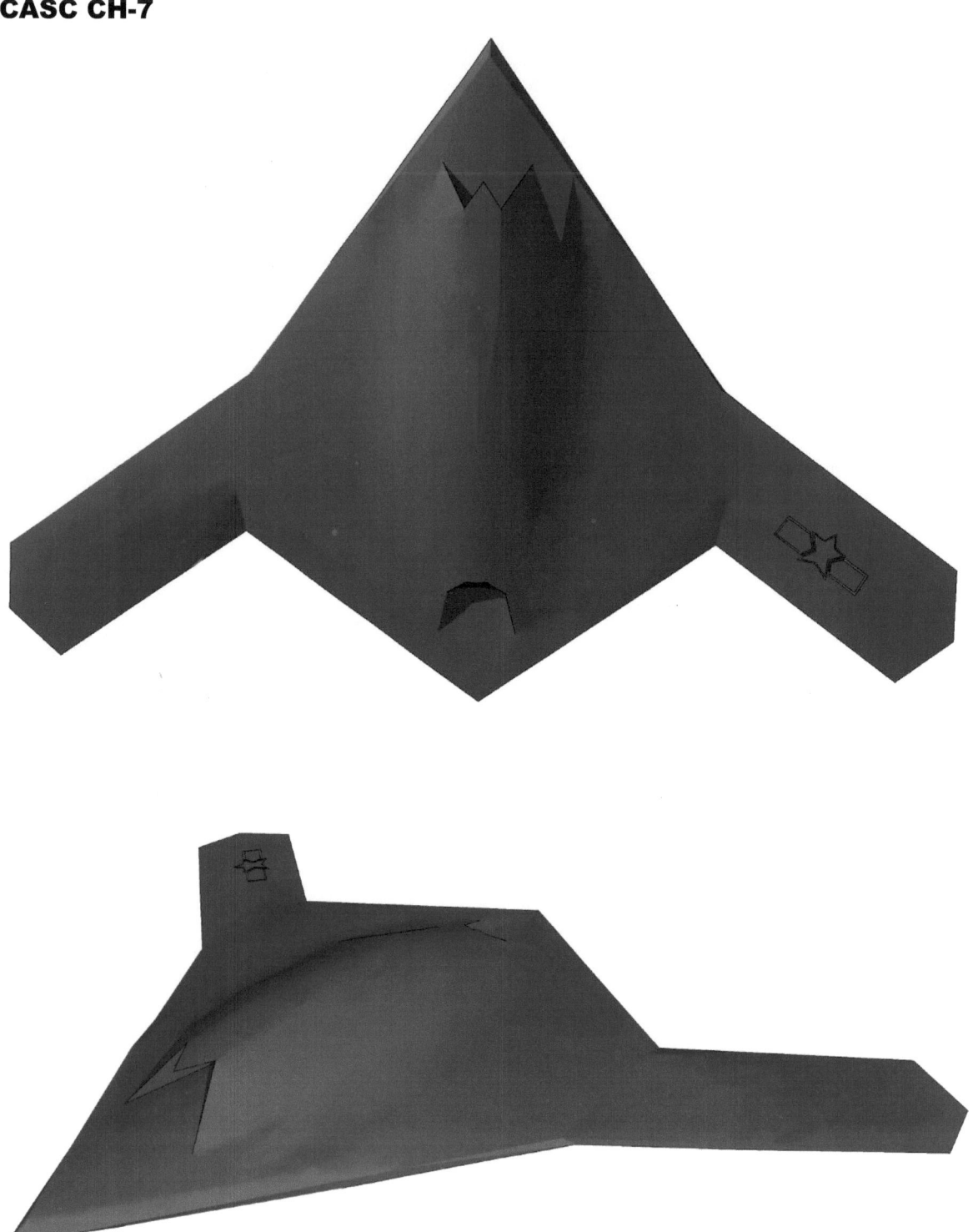

AFT J-10 Simulation Drone
Used as a training drone.

Length - 2.7 m
Empty weight - 8.5 kg
Max takeoff weight - 12 kg
Power - Jetcat P160 turbojet

AFT Single Soldier I MAV
(Dan-Bing Yi-Hao or Danbing Yihao, 单兵一号) is constructed of carbon fiber with conventional layout, powered by a two-blade propeller driven by a pusher engine installed behind the wing.

Aisheng ASN-7 Target UAV
Ba-7, (靶-7, meaning Drone-7), is the drone version of ASN-7. The B-7 is specially designed as a target for missiles, and the parachute on ASN-7 is replaced by jammers to jam the guidance of missile to simulate enemy aircraft.

Aisheng ASN-9 UAV
Conventional high fixed wing configuration, with incorporation of barometric altimeter to enable it to fly as low as 75 ft above sea level. Ba-9 (靶-9, Drone-9) was developed for maritime operations. Launched by catapult and recovered by parachute, the payload includes (2) towed targets deployed sequentially.

Length - 2.5 m
Wingspan - 2.82 m
Height - 0.72 m
Ceiling - 4 km
Remote control range - 15km
Endurance - 40 – 60 min
Max speed - 250 km/hr

Aisheng ASN-15 Reconnaissance UAV
CCD camera transmits image via datalink in a real-time. Recover by parachute or gliding.

Speed - 80 km/h
Altitude - 500 m
Endurance - 1 hour

Aisheng ASN-101 UAV
A successor of the ASN-9.

Aisheng ASN-104 Reconnaissance UAV

Equipped with four-cylinder two-stroke piston engine with air cooled HS-510 (maximum power 30 HP).
Wingspan - 4.3 m
Length - 3.32 m
Range - 60 km
Takeoff weight - 140 kg
Endurance - 2 hr

Aisheng ASN-105 Reconnaissance UAV
Similar to ASN-104 but range increased to 100 km.

Aisheng ASN-106 High Speed Target UAV
Multi-purpose, high subsonic, high mobility target drone, used as dynamic simulative targets for test of high efficiency weapon. It can simulate the electromagnetism, movement and IR feature of different kinds of targets.
Max speed - Mach 0.8
Cruise speed - 500 – 600 km/h
Minimum operating altitude - 10 m
Ceiling - 10 km
Max take-off weight - 170 kg
Payload - 20 kg
Endurance - 1 hour
Max load -7 g

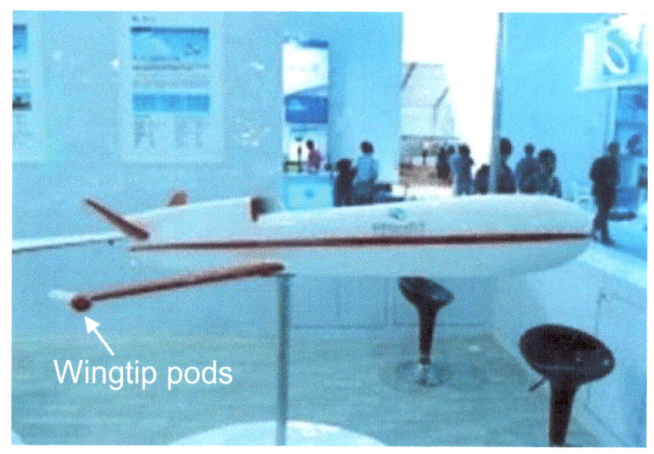

Aisheng ASN-206/207 Reconnaissance UAV
Used for day/night with electronic warfare and countermeasures (EW/ECM), battlefield surveillance, target positioning, artillery spotting, border patrol, nuclear radiation sampling, aerial photography, prospecting, and electronic countermeasures. One of most popular and tactical UAV fielded by PLA. Images are transmitted in real time.
Range - 150 km (ASN-207 = 600km)
Altitude - 19,500 m
Payload - 50 kg (ASN-207 100 kg)
Endurance - 16 hours

Aisheng ASN-209 "Silver Eagle" Multi-Purpose UCAV

Medium Altitude, Medium Endurance (MAME) tactical UAV. Two-boom, rear engine. Rocket-assisted take off, parachute recovery. Can perform aerial reconnaissance, day or night, transmits images in real time. Launched via catapult from the rear of a modified 6x6 wheeled transport truck with rocket assist. The **ASN-209H** has 3 external hard points.

Range - 200 km (radius)
Speed - 180 km/h
Altitude - 16,400 m
Take-off Weight - 320 kg
Payload - 50 kg
Endurance - 10 hours

Aisheng ASN-211 UAV

Low-altitude ornithopter (flapping wings) used for reconnaissance (200 m)
Take-off weight - 0.22 kg
Speed - 6 – 10m/s
Normal operating altitude - 20 to 200 m
Wingspan - 0.6 m
Weight - 220 g

Aisheng ASN-212 Border Patrol UAV

Small size TV, IR, digital camera or weather detection sensor. Can perform battle field survey, battle damage assessment and frontier patrol. Two engines, take-off and landing roll.

Altitude - 4,800 m
Length - 2.4 m
Wingspan - 4.2 m
Take-off weight - 50 kg
Payload - 6 – 12 kg
Speed - 60 – 90 km/hr
Endurance - 5 hr

Aisheng ASN-213 MAV

Micro air vehicle with folding wing has been linked to fast attack.
Weight - 5 kg
Payload - 1 kg

Aisheng ASN-215 Multi-Use UAV

Fixed wing UAV with conventional layout in high wing configuration. It is an enlarged version of ASN-9 and weighs twice as much. Intended for reconnaissance and battle damage assessment.

Aisheng ASN-216 Vertical Take-off and Landing UAV

Used for environment monitoring, power/oil pipeline patrol, railway patrol, surveying and mapping, border patrol. 1080p high definition image transmission. Uses infrared camera to monitor target thermal signature at night.

Range - 50 km

Aisheng ASN-217 Electric Hand-Thrown Drone

Small size TV camera or camera. Hand thrown, recovers with parachute or landing gear. Used for earthquake situation exploration, weather detection, land exploration, geography mapping, search and rescue, electricity and petroleum pipeline detection.

Range - 20 km (radius).

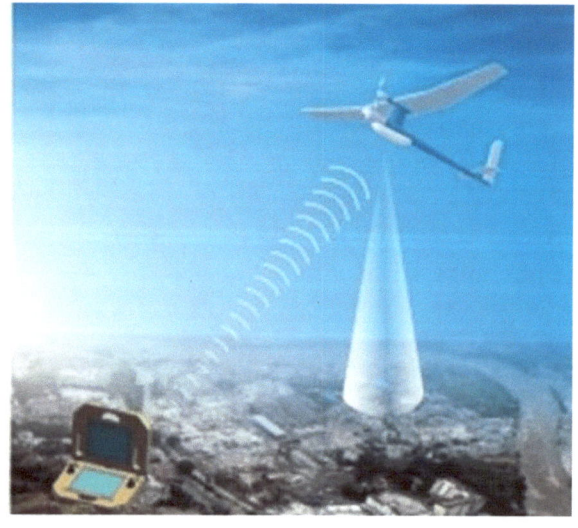

ASN-218 UAV

ASN-218 is the larger cousin of the smaller ASN-216, about twice as large, and the two UAVs look almost the same to each other, except the different tail designs.
- Speed - 140 km/hr
- Take-off weight - 40 kg
- Endurance - 4 – 8 hr

ASN-219/219A "Magpie III" Long- Endurance Reconnaissance UAV

Small and long-endurance with air pressure catapult take-off, collision rope recovery technology. Capable of carrying visible/infrared dual photoelectricity equipment, visible/infrared/laser rangefinder tri-optical optoelectronic platform, small synthetic aperture radar (SAR) and other equipment to perform real-time reconnaissance.

Range - 150 km (radius).

Aisheng ASN-229 Reconnaissance/Strike UCAV

Main task is aerial reconnaissance, electronic warfare, relay VHF radio signals and correction of artillery fire. Can carry guided missiles for precision strikes on small and moving targets. Twin boom, piston engine with two-blade pusher propeller.
Range - 2,000 km
Speed - 180 km/h (220 max)
Altitude - 10,000 m
Takeoff weight - 800 kg
Payload - 100 kg plus sensors
Endurance - 20 hours

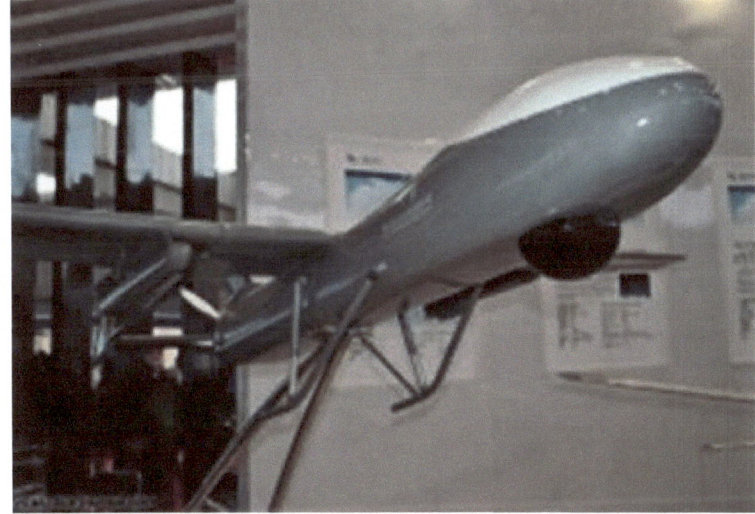

ASN-301 Anti-Radiation Radar Loitering Munition Suicide Drone

Near-copy of Israel **Harpy**. Delta-wing with pusher prop. Features passive radar seeker, used as a long-range anti-radiation weapon to attack radar stations. Newer launch vehicle can carry 6 box launchers.

Range - 500 km
Speed - 185 km/hr
Warhead - 32 kg

Aisheng BZK-600 UCAV

Evolved from the ASN-205/207 (WZ-6, but larger). Ground attack and recon roles. Rocket-assist launch from flatbed of a Chinese Army truck.
(4) air-to-surface missiles
Length - 4.3 m (14.1 ft)
Endurance - 12 hr

Aisheng DCK-006 Reconnaissance UAV

Unarmed reconnaissance version of BZK-006

Aisheng JWP02 Reconnaissance UAV

Altitude - 5,000 m
Length - 4.273 m
Wingspan - 7.5 m
Takeoff weight - 320 kg
Payload - 50 kg
Endurance - 10 hr

ASN C-31 VTOL UAV

Aisheng WZ-8 High-Speed, High-Altitude Reconnaissance Drone

Also called the **DR-8**. Air-launched drone powered by two small open-cycle, liquid-fueled, rocket engines.
Range ~ 1,500 miles
Altitude ~ Less than 140,000 ft
Speed ~ Mach 3

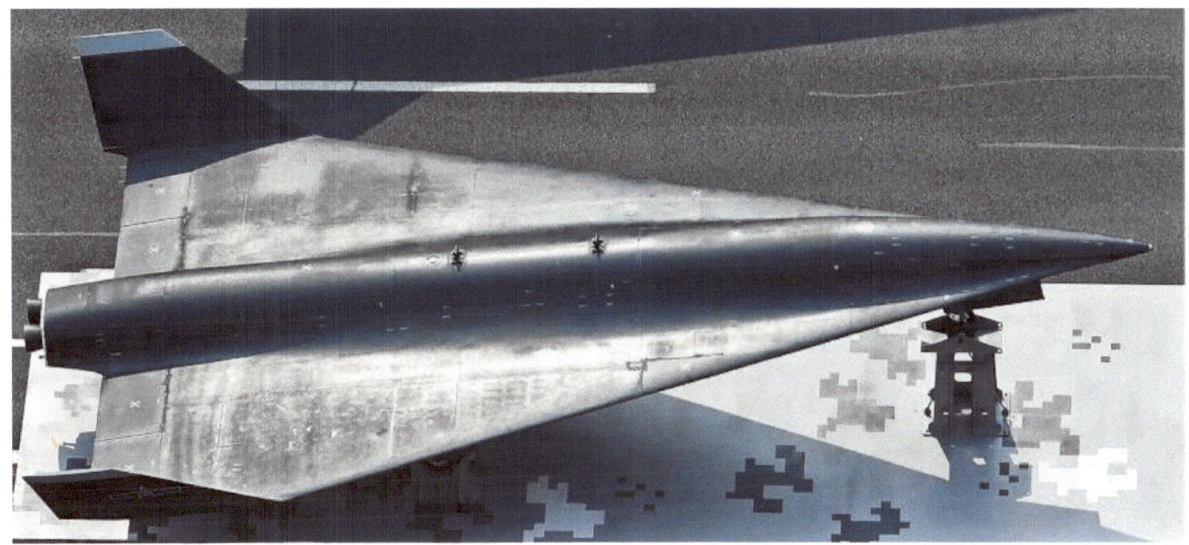

AT200 Cargo Drone

Unmanned cargo aircraft features low-wing monoplane design with a large fuselage.
Range - 2,100 miles
Altitude - 6,000 m
Speed - 313 km/h
Length - 11.84 m
Wingspan - 12.8 m
Max. take-off weight - 3.4 t

AVIC Yun Ying "Cloud Shadow" HALE Reconnaissance/Strike

The UAV wing has six pilons. Can be equipped with 50 kg CS/BBM3 (YL-12) GPS guided bomb, Blue Arrow air-to-surface missile, 100 kg GB-4 precision-guided bomb, and light cruise missiles.
   Cruise speed - 620 km/h (390 mph, 330 kn)
   Altitude - 15,000 m (49,000 ft)
   Length - 9 m (29 ft 6 in)
   Wingspan - 20 m (65 ft 7 in)
   Height - 3.66 m (12 ft 0 in)
   Empty weight - 2,300 kg (5,071 lb)
   Max takeoff weight - 3,200 kg (7,055 lb)
   Powerplant - WP-11C turbojet
   Endurance - 6 hours

AVIC 601-S "Cloud Bow" UAV

(Yun-Gong or Yungong, 云弓) experimental UAV, the winglets on earlier Wind Blade is eliminated, and hence Cloud Bow is also called 4 control surfaces flying wing. (si duo-mian fei-yi, 四舵面飞翼).

AVIC 601-S "Dark Sword" Air-to-Air UCAV

AKA "Anjian"
Limited info available

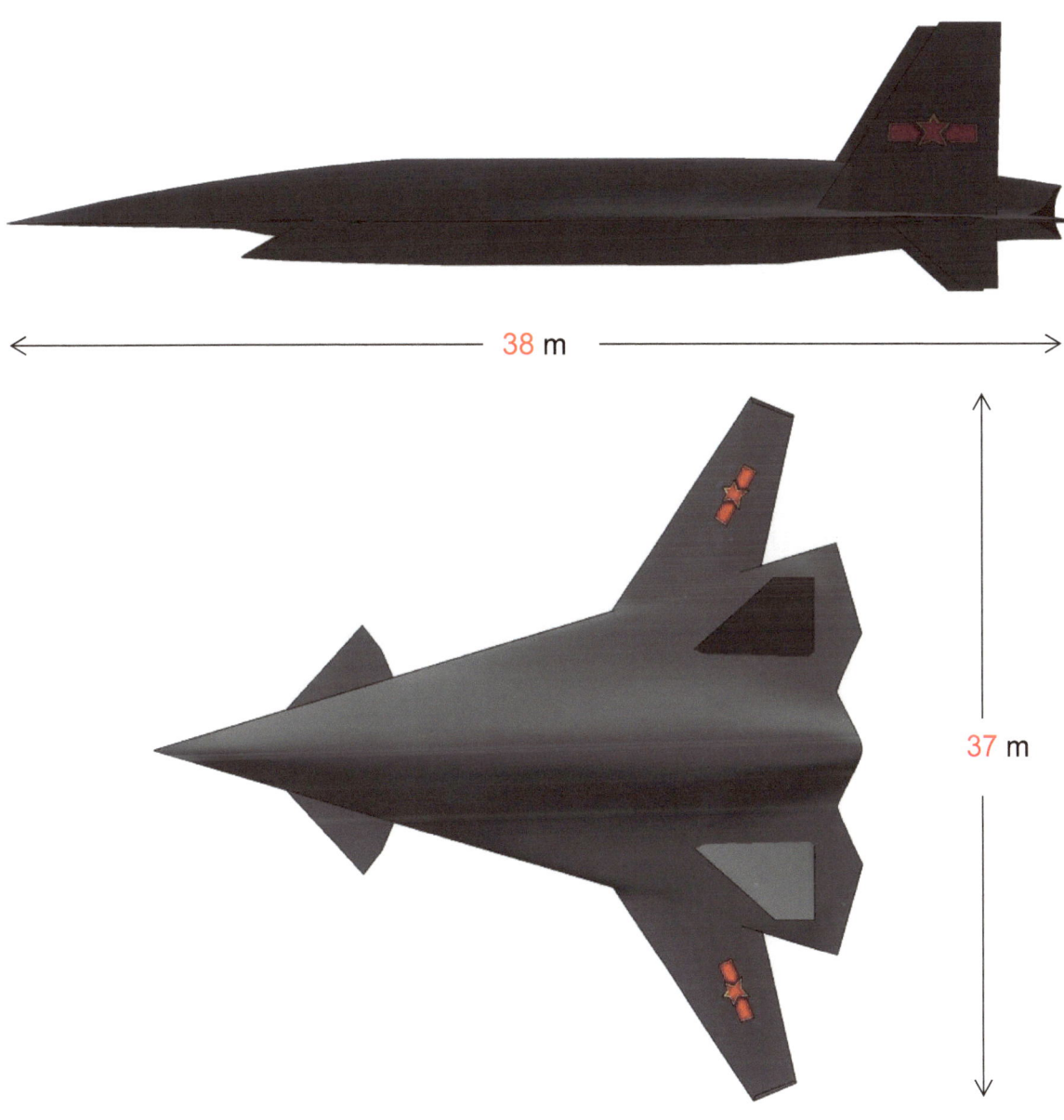

AVIC 601-S "Sky Crossbow" UCAV

Experimental UAV with internal propeller flying wing with V-tail.

Length - 2.15 m
Wingspan - 2.2 m
Height - 0.6 m
Weight - 18.8 kg
Speed - 70–150 km/h
Power plant - Electrically powered ducted fan

AVIC 601-S "Warrior Eagle"

Experimental with forward swept flying wing design.

AVIC 601-S "Wind Blade" Drone

(Feng-Ren or Fengren, 风刃) The twin-tail is replaced by winglets.

AVIC AW-4 "Shark II"

Shark II (Sha Yu or Shayu 鲨鱼) is a light UAV of conventional layout with V-tail, and propulsion comes from a two-blade propeller driven by a pusher engine.

Range - 150 km
Cruise speed - 100 km/h
Ceiling - 4 km
Max takeoff weight - 55 kg
Endurance - 4 hours

AVIC L-15 "Blue Fox" Target Drone

Capable of simulating the modern, high maneuverability fighters from third generation up. The target drone will be powered by two turbojets, each providing maximum thrust of 60dN.

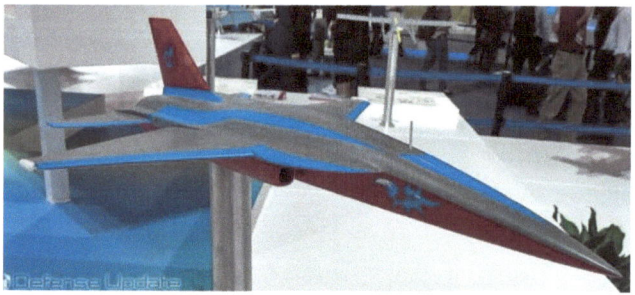

AVIC Short-Tailed Falcon Drone

Model shown

AVIC "Sky Eye" Disposable Artillery Drone

The Sky Eye is a one-time "artillery drone" currently in service. The delivery vehicle for this flying spotter is a rocket launcher system. It can also be placed in 155 mm self-propelled howitzer PLZ-04. The projectile flies along a ballistic trajectory, and according to the timer signal, at a given point, it opens and brakes with a parachute. When the speed drops to the minimum value, the drone detaches from the projectile and expands the rotor blades rotated by an electric motor. The device freezes at a certain height and with the help of a television camera begins to search for a target which it then illuminates with a laser.

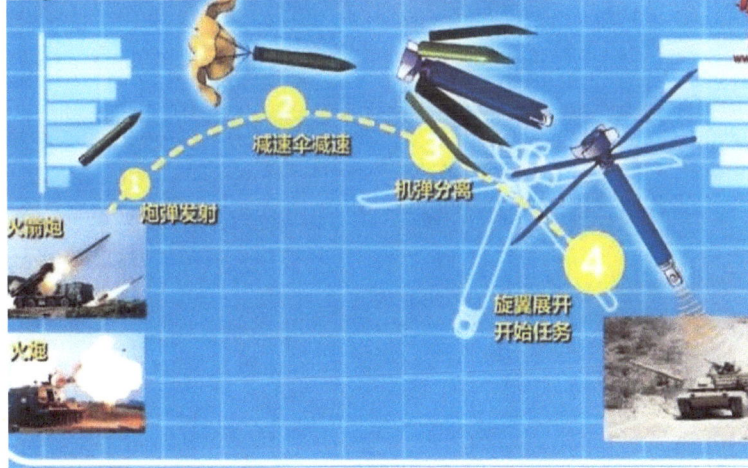

AVIC TL-8 "Sky Dragon" Target Drone

TL-8 (TL= Tian Long, 天龙 Sky Dragon) was developed to simulate a Tomahawk missile cruise missile (size, shape, speed, flight characteristics).

AVIC "Whirlwind Scout" MAV

The ducted-fan design, is similar to Honeywell's and is able to remain airborne for 20 to 40 minutes and follow a flight path of up to 100 waypoints. With noise of around 60 dBA, the system cannot be audibly detected by a human for 560 m. Target location error is 40 mm. Communications range is 5 to 10 km.

Speed - 90 km/hr
Altitude - 3,000 m
Takeoff weight - 8 kg
Payload - 1.5 kg

AVIC XLB "Patroller" UAV

Powered by a fuel cell.
Data Transmission Range - >50 km
Speed - 150k m/h
Ceiling - 5,000 m
Length - 4.7 m
Wingspan - 10.5 m
Payload - 30 kg
Endurance - 3-5 hours

AVIC YY-1 "Swift" MAV

YY-1 Swift has a conventional layout with an auxiliary lift propeller engine at the base of its V-tail, and two propellers installed on the outer half of NACA 4415 wing, and this portion of the wing will be tilted during takeoff and landing.

BESTUAV SY-5 "Divine Eagle" 5 HALE UCAV

Very large twin-fuselage UCAV designed for operational reconnaissance. Long range radar would allow it to stay away from air defenses like the AEGIS. Divine Eagle could datalink with SAMs on land and sea to shoot down low-flying cruise missiles.

   Ceiling - 4.5 km
   Weight - 120 kg
   Payload - 20 kg
   Endurance - 5 hr

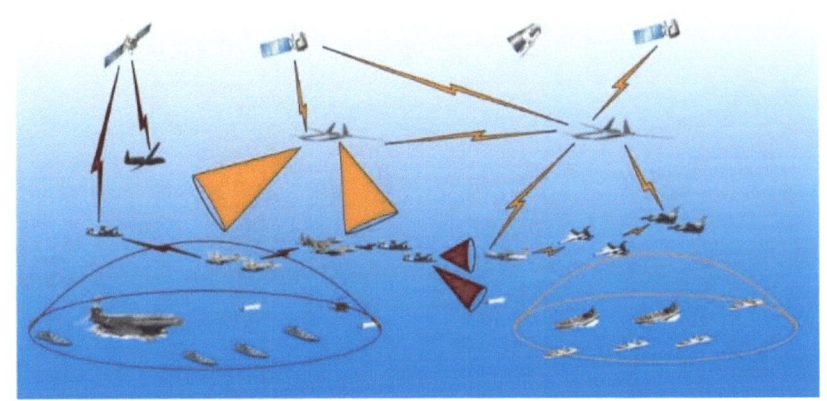

BIT "Falcon" Experimental Thrust-Vectoring UAV

Falcon (Lie-Ying, 猎鹰) is an experimental fixed wing UAV intended to explore technologies of thrust vectoring.

BIT Gun-Launched UAV

This electrically powered UAV is deployed by a 155 mm round, with a max payload of 2.9 kg

Blowfish A2 Helicopter Drone -

Belly-less fuselage, can carry radar, jamming devices, guns or bombs under its spine. Built by Guangdong-based Zhuhai Ziyan UAV Co.

Speed - 130 km/h
Takeoff weight - 38 kg
Payload - 12 kg

BMP LHK

Twin-boom with inverted v-tail, two-blade propeller pusher 55 cc engine mounted at the rear with tricycle landing gear.

- Speed - 110–150 km/h
- Altitude - 5 km
- Length - 3 m
- Wingspan - 2.53 m
- Takeoff weight - 25 kg
- Payload - 10 kg
- Endurance - 3.5 h

BMP YZ-8 UAV

YZ-8 is a large fixed UAV of YZ series, and it's in conventional layout with high wing configuration and V-tail. Propulsion is provided by a two-blade propeller driven pusher engine. Winglets are used and tricycle landing gear. YZ-8 has modular design and subsystems such as engine can be changed rapidly among different types, and fuel tanks can be added or reduced based on mission requirement. Externally, YZ-8 looks very similar to the GAIC Harrier Hawk II Air Sniper. The YZ-8 propeller of the former only has two blades and the Harrier Hawk II has three. There is a bulge on the top of the nose of YZ-8 housing communication satellite dish, which the fuselage of Harrier Hawk II Air Sniper does not. Also, the YZ-8 has winglets while the Harrier Hawk II Air Sniper does not.

BUAA FH-1

Range - 90 km
Max speed - 100 km/h
Altitude - 2.5 km
Main rotor diameter - 2.6 m
Height - 1.3 m
Length - 0.8 m
Max takeoff weight - 90 kg
Payload + fuel - 40 kg
Endurance - 2 hr

BUAA Logistics Unmanned Cargo Aircraft

Range - 1,500 km
Max level speed - 400 km/h
Cruising speed - 360 km/h
Effective cargo capacity - 15.7 m
Takeoff Weight - 3.600 kg
Max payload - 1,200 kg

BZK-005 "Giant Eagle" MALE/HALE UCAV

BZK-005E or Chang Ying is comparable to the U.S. Global Hawk. Designed for long-range reconnaissance and electronic intelligence (ELINT) missions with active/passive all-weather capability (including collecting of passive radio signals). Very high lift-to-drag ratio and modular design. Based on the Predator, built by Harbin, also called "**Sea Eagle**" or "**Giant Eagle**". It has a large wingspan with twin tail booms and V-shaped tailfins constructed using composite materials. It has a three-blade propeller, pusher piston engine and is automatically controlled with artificial intelligence so that takeoff and monitoring are controlled by one button and it flies and lands automatically. Artificial intelligence is also used for navigation, flight control and performance of tasks. It can search and track its targets automatically and identify the position of targets in real time. It can carry 650 kg of fuel and equipment including photoelectric, radar and electronic reconnaissance systems and electronic warfare equipment. Turret installed underneath the nose houses the FLIR/CCD cameras using integrated super long-distance imaging system that integrates various sensors of various frequencies of light. At altitude of 3,000 meters, it can clearly read the number of the license plate of a vehicle on the ground. The variant in service with PLA Army ("**Sky Eagle**") features a chin-mounted SAR antenna with the EO turret relocated to the mid-section of the fuselage. The BZK-005C features 4 underwing pylons and is able to carry up to 300 kg of various weapons. A strike-capable variant called **TYW-1** can carry a 370 kg payload and features four underwing pylons for carrying air-to-air, air-to-surface missile, or ISR pods. It is believed to have a maximum take-off weight of 1,500 kg compared with the 1,250 kg of the BZK-005.

Range - 2,400 km
Speed - 150-180 km/h
Altitude - 8,000 m
Takeoff weight
 BZK-005 = 1,250 kg
 TYW-1 = 1,500 kg
Payload
 BZK-005 = over 150 kg;
 BZK-005C = 300 kg
Endurance - 40+ hours

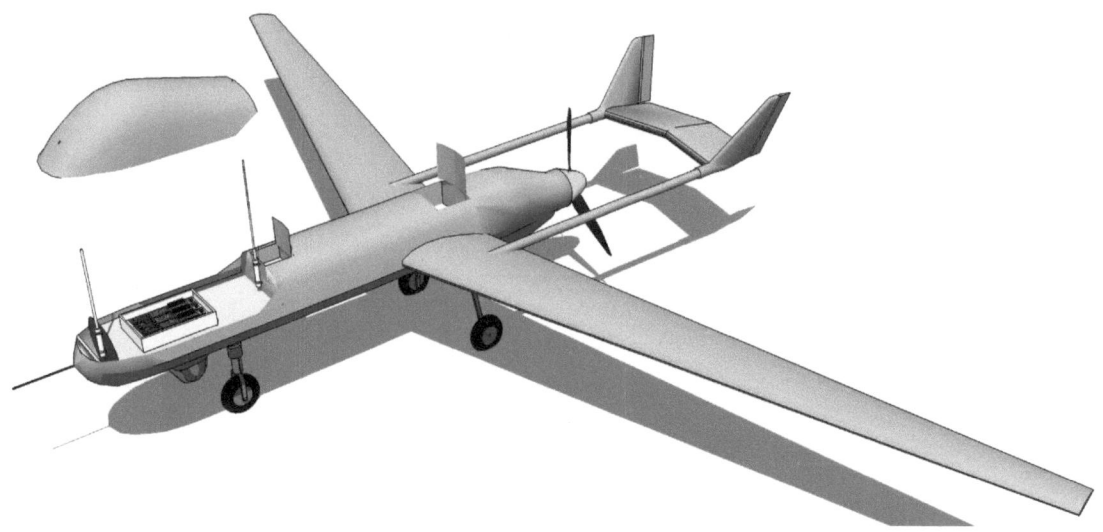

CADI "Nimble Loong" MALE Short-Range UAV

Similar to the Scan Eagle and RQ-21A Blackjack. Launched from a catapult and retrieved by suspended cable arrest.

Endurance - 6 – 8 hr

CAIG "Sky Wing" I UAV

  Range - 100 km
  Max speed - 180 km/h
  Cruise speed - 120–150 km/h
  Ceiling - 3 km
  Max takeoff weight - 80 kg
  Payload - 20 kg
  Endurance - 3 h

CAIG "Sky Wing III" HALE UAV

Appears similar to Northrop Grumman RQ-4 Global Hawk. Sky Wing III is equipped with a jet engine mounted on top of the fuselage between the V-shape tail wings.
Range - 7,000 km
Speed - 750 km/h
Ceiling - 15 km
Length - 8.9 m
Height - 3.44 m
Wingspan - 18.8 m
Weight - 7.5–9.1 t
Payload - 0.65–2+ t
Endurance - > 6 hr

CAIG GJ-I UCAV

The GJ-1 is a land-attack version of Pterodactyl I that combines the capabilities of both Pterodactyl I and WJ-1 so that it can identify and engage targets on its own. GJ-1 can be distinguished from both Pterodactyl I and WJ-1 in that GJ-1 has both the reconnaissance/targeting pod under the chin as well as hardpoints to carry weapons. The designation GJ stands for Gong-Ji Wu-Ren-Ji (攻击无人机), meaning "attack UAV" or "attack drone".

CAIG Wing Loong I "Pterodactyl I" MALE UCAV

The CAIG Wing Loong I (Chinese: 翼龙-1; pinyin: Yìlóng-1, '**Pterodactyl**'), based on the Raptor, can carry the BA-7 air-to-ground missile, YZ-212 laser-guided bomb, YZ-102A anti-personnel bomb and 50-kilogram LS-6 miniature guided bomb.

Performance
Range - 4,000 km (2,500 mi)
Speed - 280 km/h (170 mph)
Altitude - 5,000 m (16,000 ft)
Payload - 1,000 kg (2,200 lbs.)
Endurance - 20 hr

Armaments
Bombs
FT-10, FT-9/50, FT-7,
GB7, GB4
Missiles
BRM1, AKD-10

Avionics
100 kilograms
(220 lbs.) capacity for
sensors

CAIG Wing Loong II "Pterodactyl II" MALE UCAV

CAIG Wing Loong II (Chinese: 翼龙 II; pinyin: Yìlóng Èr; '**Pterodactyl II**', military designation **GJ-2**) is capable of remotely controlled or autonomous flight. It is an enlarged version of Wing Loong I with longer body and wider wing span. It has a slender fuselage, V-tail and ventral fin. It has retractable landing gear, including two main wheels under the fuselage and one single wheel under the nose. Each wing has three hardpoints for up to twelve weapons.

CASC CH-3 Fixed Wing UCAV

Unusual canard layout, 3-blade prop, pusher engine.

Range - 960 km
Altitude - 4 km
Payload - 60-80 kg
Endurance - 12 hr

CASC CH-3A Reconnaissance/ Strike UCAV

(Cai Hong; '**Rainbow**') with satellite data link can carry AR 1 laser-guided missile.

Range - 960 km
Altitude - 4 km
Payload - 180 kg
Endurance - 6 hr

CASC CH-4 Reconnaissance/Strike UCAV

Caihong-4 (CH-4) armed drone is based on stolen plans for the General Atomics MQ-9 Reaper with fewer outboard stations for mounted ordnance.

Range - 4,000 km
Speed - 345 km/h
Altitude - 22,960 m
Payload - 345 kg
Endurance - 40 hr

CASC CH-5 "Rainbow 5" UCAV

Based on the CH-4.

Range - 2,000 km
Speed - 1,200 km/h
Altitude - 7,000 m
Endurance - 40 hr

CASC CH-91 Fixed-Wing Reconnaissance and Surveillance UAV

Twin-boom layout with inverted v-tail and a pair of skids as landing gear and rocket-assisted takeoff. AKA BZK-008.

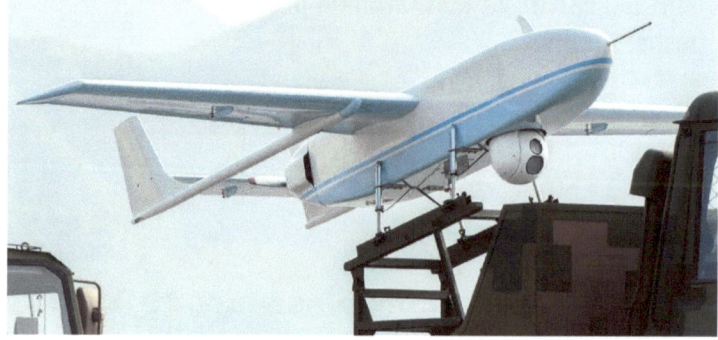

CASC CH-92 Fixed-Wing Reconnaissance and Surveillance UAV

Conventional layout with V-tail and tricycle landing gear. Designed to conduct large-area observation and precision strikes, to be launched by a spring system or rocket accelerator, to be recovered by a parachute system or landing arresting device.

CASC CH-802 Small Hand-Thrown Launch Reconnaissance and Surveillance UAV

Conventional layout with elevated high-wing and V-tail. 2-blade propeller.

Range - 30 km (radius)
Speed - 60 km/h
Altitude - 4 km
Payload - 1 kg
Endurance - 4.5 hr

CASC CH-803 Fixed-Wing Reconnaissance and Surveillance UAV

Cylindrical fuselage and canards, without tailplane.
Range - 30 km (radius)
Speed - 80-110 km/h
Altitude - 3.5 km
Payload - 3.5 kg
Endurance - 5 hr

CASC CH-805 Training UAV

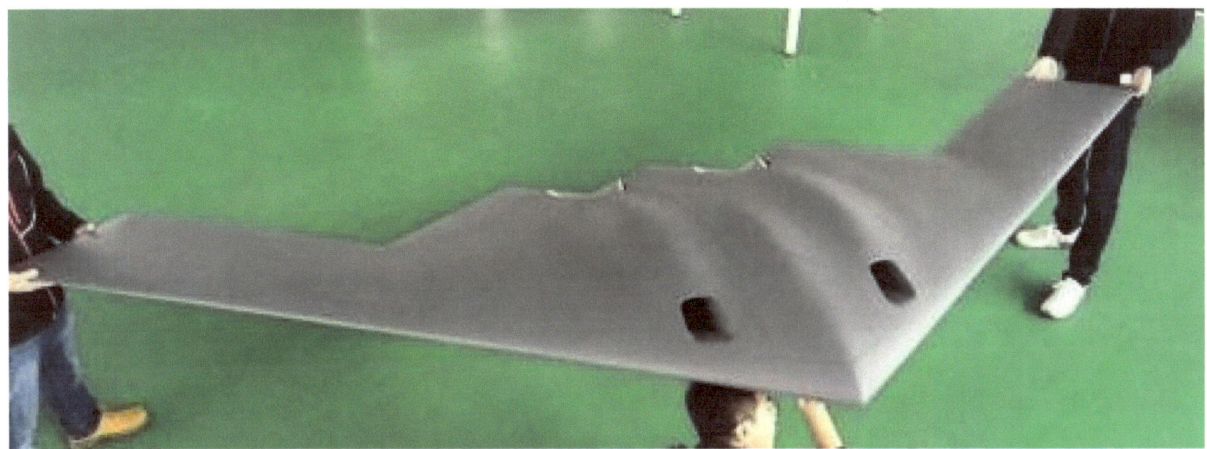

It is a small drone, launched with a booster that uses a rocket engine.
Speed - 730 km/h
Take-off weight - 190 kg
Endurance - 40 minutes
Wingspan - 4 m

CH-806 Small Long-Endurance Reconnaissance and Surveillance UAV

Both visible/infrared payloads and miniature synthetic aperture radars can be mounted according to mission requirements. Can complete the flight control and mission planning and load, link monitoring and information processing. Reconnaissance images can be received at multiple points.

CASC CH-901/BG-201 Kamikaze Drone UCAV

Cylindrical fuselage and high-wing, two-blade propeller driven by pusher engine. Combines the advantages of a drone and a bomb. Launched from the field to track down and attack enemy. Explosive warhead can penetrate light armored vehicle (with a fragmentation charge or a shaped charge). Able to detect target at 1.5 km.

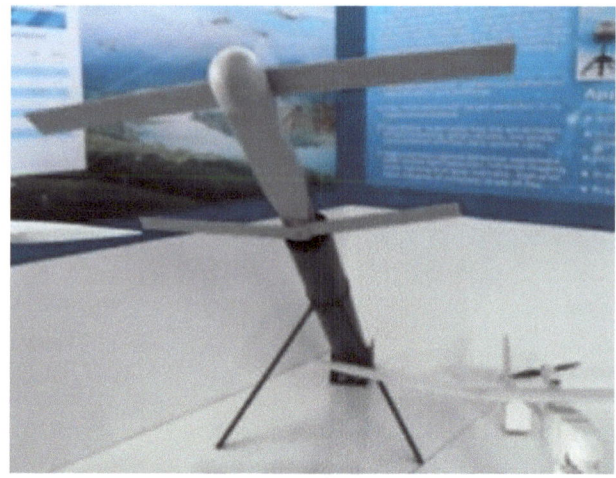

Range - 15 km
Speed - 7 to 150 km/h
Altitude - 450 m
Weight - 9 kg
Endurance - 40 minutes

CASC "Peace Map" UAV

A unique feature of Peace Map UAV is that it has triple tails.

CASC PW-2 Medium and Short-Range UAV

Mainly for battlefield reconnaissance, correction of artillery fire, data relay, electronic warfare and intelligence collection.

CASC SN-3A UCAV

UCAV has satellite guidance. In addition, as the payload can be installed radar with synthetic aperture antenna, electronic warfare equipment and apparatus relaying radio signals.

Range - 2,000 km
Radius - 200 km
Altitude - 5 km
Cruising speed - 180 km/h

Maximum speed - 240 km/h
Length - 5.1 m
Wingspan - 7.9 m
Height - 2.4 m
Takeoff weight - 640 kg
Payload - 100 kg
Duration is 12 hours.

Weapons
AR-1 laser-guided missile (mass 45 kg)
FT-25 compact controlled bomb (weight 25 kg)
Can carry two bombs FT-5 caliber 75 kg (warhead - 35 kg, QUO – 3-5 m)

CASC CH-T4 Solar Drone

CH-T4, the second largest solar drone in the world.
Altitude - 20,000 m

CASIC HW-600 Reconnaissance/ Strike UCAV

Reconnaissance (WJ-600) and ground attack (HW-610) variants.
Note: MANPADs can be deployed by this UCAV as air-to-air missiles.

Range - 2100 km
Speed - 200 m/s
Altitude - 10 km
Launch - rocket-assist
Recovery - parachute
Payload - 130 kg
Endurance - 3 - 5 hr

CASIC HW-610 Reconnaissance/ Strike UCAV

Speed - 500 – 700 km/hr
Length - 6.5 m
Wingspan - 6.8 m
Take-off weight - 1 t
Payload - 130 kg
Endurance - 3 – 5 hr
Remote control radius - 350 km
Launch - Rocket assisted take-off
Recovery - parachute

CASIC HW-800

The only information about HW-800 confirmed is that it is a fixed wing UAV.

CASIC HW-X100 "Soar Cloud"

HW-X100 Soar Cloud (Xiang-Yun or Xiangyun, 翔云) UAV. Very little is known other than that it is currently under development by HiWING.

CASIC HW-X200

HW-X200 is another very little-known UAV currently under development by HiWING.

CASIC SF-460 Blade

A derivative of HW-300, with identical layout except the SF-460 has winglets, and SF-460 is larger than HW-300.

Speed - 110 km/h
Altitude - 3 km
Length - 3.3 m
Wingspan - 4.6 m
Weight - 85 kg
Payload - 15 kg
Endurance: 4–6 h
Power plant - 13 kW engine

CASIC "Sky Hawk 1" UCAV

Length - 2.3 m
Wingspan - 4.2 m
Altitude - 5 km
Speed - 90–150 km/h
Combat radius - > 180 km
Max takeoff weight - 150 kg
Payload - 30 kg
Endurance - 6 h

CASIC "Spiderman" Fighting Drone

Fires a 172 sq. ft net over target drone.

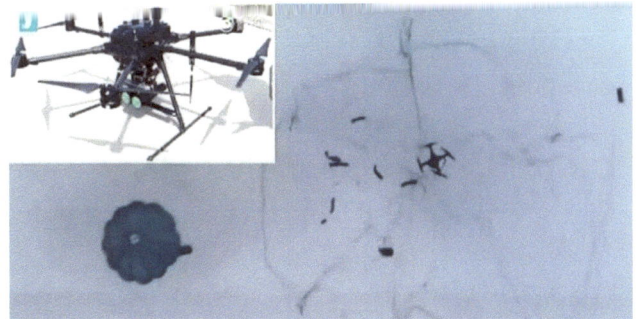

CASIC TX-1 UAV

Dubbed as flying saucer UAV (Die Xing Fei-Xing-Qi, 碟形飞行器), is an experimental UAV powered by two jet engines mounted atop of the fuselage at the root of the single vertical rudder.

CASIC WJ-500 UCAV

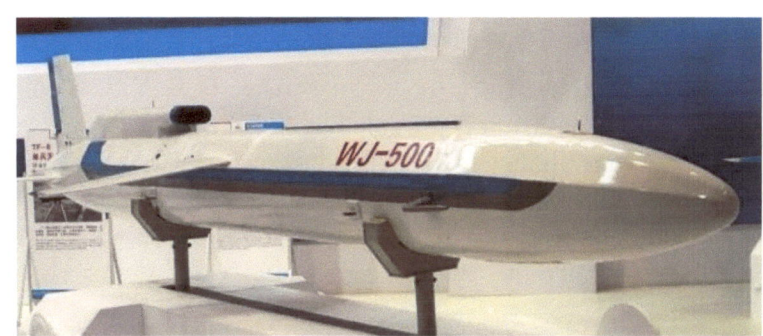

CASIC WJ-600 A/D UCAV

See CASIC HW-600

CASIC WJ-700 UCAV

Latest model can carry CM-102 anti-radiation missile, the C701 ground attack missile, and the C705KD anti-ship missile. It can also operate as a decoy, simulating targets such as cruise missiles and aircraft

Takeoff weight - 3,500 kg
Endurance - 20 hr

CAUC XC-1 "Flying Shuttle" Submarine Launched UAV

XC-1 Flying Shuttle (Fei-Suo or Feisuo, 飞梭) is a very little-known experimental UAV developed by CAUC to study feasibilities of submarine launched UAVs.

CDADI Coaxial Shape-Varying Rotary Wing Artillery-Launched UAV

Can be either used as a reconnaissance platform, or a cruise missile to attack selected targets.

CDADI "Mad Warrior" Shape-Changing Wing UAV

Experimental UAV that changes the shape of its wing at different speed for better performance. At supersonic speed, the outer portion of the main wing folds upward to a ninety-degree position perpendicular to the inner portion of the main wing and canards completely retract into the fuselage. At subsonic speed, canards are fully extended and the out portion of the main wing would not be folded.

CDADI "Sea Patroller" Ocean Surveillance UAV

Sea Patroller (Hai-Xun-Zhe or Haixunzhe, 海巡者) UAV is a fixed wing UAV is twin boom layout, and propulsion is provided by a two-blade propeller driven pusher engine. Designed for ocean surveillance missions and thus carries a various electro-optical equipment.
   Range - 200 km
   Endurance - 4, 6 or 8 hr
   Launch – catapult

CDADI VD200 UAV

A tail-sitter in flying wing layout with two vertical stabilizers. Propulsion is provided by a two three-blade propeller-driven engines with one mounted on each wing. After taking off, the entire fuselage would pitch 90 degrees and flies like a normal fixed wing UAV. Similar to Lockheed "Fury".

CH-500 Dual-Rotor Helicopter Drone

CH-500 UAV with Blue Arrow-9 (BA-9) Air-to-Surface Missile.
Range - 6 km
Takeoff Weight - 26.5 kg

CH-805 Stealth Drone

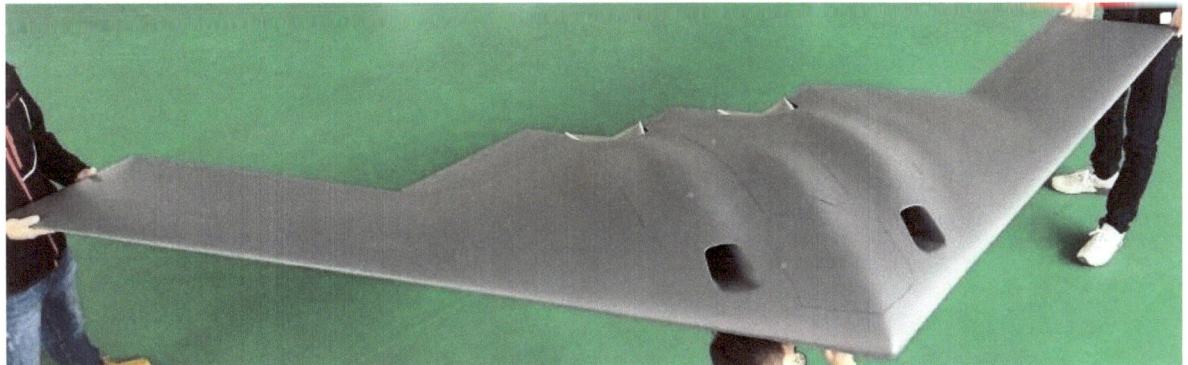

Speed - 730 km/h
Wingspan – 4m
Takeoff Weight - 190 kg
Endurance - 40 minutes

GAIC BZK-007 UCAV

The BZK-007 has a fixed tricycle landing gear. PLAAF use it as tactical reconnaissance UAV carrying variety of equipment including FLIR, CCD TV cameras and remote sensors.

Wingspan - 14.6 m
Height - 2.74 m
Length - 7.7 m
Maximum takeoff weight - 700 kg
Max speed - 230 km/h
Payload - 100 kg max
Altitude - 7.5 km
Endurance - 16 h

GAIC "Guizhou Central" UCAV

Guizhou Central (Qian-Zhong or Qianzhong, 黔中) UAV is another derivative of Sunshine UAV developed by GAIC. The most distinct external visual difference between Guizhou Central and BZK-007/Sunshine is that the fuselage of Guizhou Central is more aerodynamically refined, shaped like a shuttle. The elimination of the bulk in the fuselage means the elimination of satellite dish antenna for long range communications. Guizhou Central is equipped with a retractable tricycle landing gear system, and like BZK-007 and Sunshine, it is also equipped with a three-blade propeller driven by a tractor engine mounted in the nose.

Wingspan - 14 m
Height - 2.6 m
Length - 7.7 m
Weight - 660 kg

Max speed - 160 – 180 km/h
Payload - 70 kg max
Altitude - 7.5 km
Typical radius - 230 km
Endurance - 15 h

GAIC "Harrier Hawk II Air Sniper" UCAV

GAIC "Harrier Hawk III" UAV

This is a flying-wing / blended wing body layout. Propulsion is provided by a pair of three-blade propellers driven by a pusher engine mounted at the rear end of the flying wing. Harrier Hawk III has retractable tricycle landing gear system and winglets.

GAIC "Sunshine" Reconnaissance UAV

Sunshine (Yang-Guang or Yangguang, 阳光) UAV is a fixed-wing UAV in conventional layout developed by GAIC and has entered service with Chinese authorities mainly for remote sensing. There are several features that distinguish Sunshine from the rest of Chinese UAVs; Sunshine adopts a low wing configuration and a retractable tricycle landing gear system. Sunshine is also equipped with a satellite dish communication antenna housed in the bulge in the fuselage, where the cockpit of manned aircraft would be located, so that information such as data and images can be passed down to a command and control center hundreds or thousands of miles away in near real time. Sunshine is powered by a three-blade propeller driven by a tractor engine mounted in the nose, as opposed to most Chinese UAVs that are equipped with a two-blade propeller driven by an engine.

- Wingspan - 14.6 m
- Height - 2.74 m
- Length - 7.7 m
- Weight - 750 kg
- Max speed - 240 km/h
- Payload - > 70 kg

GJ-1 MALE UCAV

Yilong drone is capable of strike role with: satellite-guided FT-7 aerial bomb, BRM-1 precision guided missile, SDBI/II small-diameter bomb.

Altitude - 5,300 m
Range - 4,000 km
Weight - 1.1 ton

GJ-2 MALE UCAV

PLA version of Wing Loong II, with synthetic aperture radar and an electro-optical pod, six pylons carry up to 12 small missiles, incl KD-9/10 laser- guided anti-tank missile (ATGM).

Speed - 370 km/h
Altitude - 9,000 m
Endurance - 20 hr

WZ-2000 Reconnaissance/Strike UCAV

The Guizhou WZ-2000, also known as the WuZhen-2000 or WZ-9. Onboard remote sensors include a thermal imaging camera and synthetic aperture radar (SAR). Imagery is transmitted via satellite communications antenna mounted in the nose bulge. The WZ-2000B is an improved production variant.
Range - 2,400 km (1,500 mi, 1,300 nmi)
Combat range - 800 km (500 mi, 430 nmi)
Service ceiling - 18,000 m (59,000 ft)
Maximum speed: 800 km/h (500 mph, 430 kn)
Payload - 80 kg (176 lbs.)
Length - 7.5 m (24 ft 7 in)
Wingspan - 9.8 m (32 ft 2 in)
Max takeoff weight - 1,700 kg (3,748 lbs.)
Endurance - 3 hours
(2) KD2 air-to-surface missiles
Additional number of ZD1 precision-guided bombs

HAIG "Blue Fox" UAV

Blue Fox (Lan-Hu or Lanhu, 蓝狐) is a subsonic UAV was first revealed at the 10th Zhuhai Airshow in 2012. Externally, Blue Fox resembles a scaled down Hongdu L-15, and is powered by a pair of turbojet engines, each providing 60 kN thrust.

    Max speed - 750 km/h
    Min speed - 10 km/h
    Altitude - 8 km
    Endurance - > 40 min
    Maneuver overload - -2 to +6

HAIG "Nighthawk" UCAV

Nightjar (Ye-Ying or Yeying, 夜鹰) is a UAV developed by HAIG with applications in both civilian and military. The aircraft is a twin-boom configuration with inverted v-tail, and propulsion is provided by a two-blade pusher engine. The landing gear system is a pair of skids and the typical payload is an electro-optical pod below the fuselage.

    Max speed - 170 km/h
    Range - 120 km
    Endurance - 3 h
    Altitude - 3.5 km

Hongdu GJ-11 "Sharp Sword"

The Shenyang Hongdu GJ-11 "Sharp Sword" (Chinese: 利剑; pinyin: Lì jiàn)

    Range - 4,000 km (2,485 mi, 2,160 nmi)
    Altitude - 12,500 m (41,010 ft)
    Cruise speed - 1,000 km/h (621 mph, 540 kn)
    Length - 11.65 m (38 ft 2.6 in)
    Wingspan - 14 m (45 ft 11 in)
    Height - 3.1 m (10 ft 2.0 in)
    Empty weight - 6,350 kg (13,999 lb)
    Max takeoff weight - 20,215 kg (44,566 lb)
    Powerplant - (1) turbofan

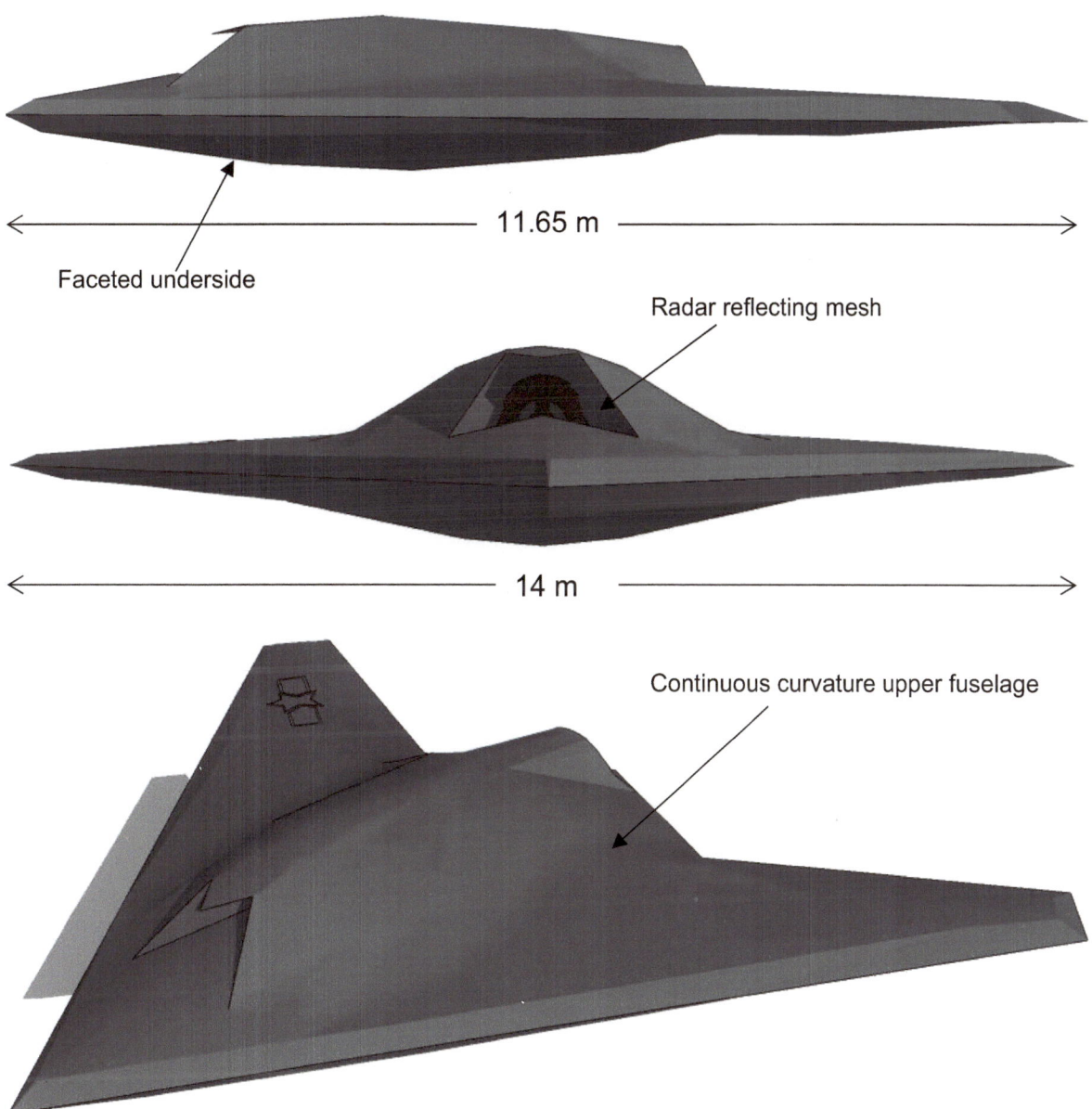

Keyuan AD200 "Blue Eagle" UAV

AD 200W Blue Eagle (Lan-Ying or Lanying, 蓝鹰) is a twin-boom configuration with canards added to the nose. The twin tails of AD200W are joined at the top, instead of at the root or midpoint as most twin tail configuration. AD200W Blue Eagle can be assembled in the field within forty minutes for rapid deployment.
Range -– 2,000 km
Speed - 135 km/hr
Max takeoff weight - 750 kg
Max payload - 300 kg
Endurance - 12 hr

LGAA LN60F "Thunderbird" UAV

LN60F Thunderbird (Lei-Niao 雷鸟) is an UAV powered by hydrogen fuel cells.
   Wingspan - 10.5 m
   Length - 4.7 m
   Height - 2.2 m
   Max takeoff weight - 257 kg
   Speed - 120 km/h
   Endurance - 4 h
   Payload - > 100 kg

Linkall HK Drone

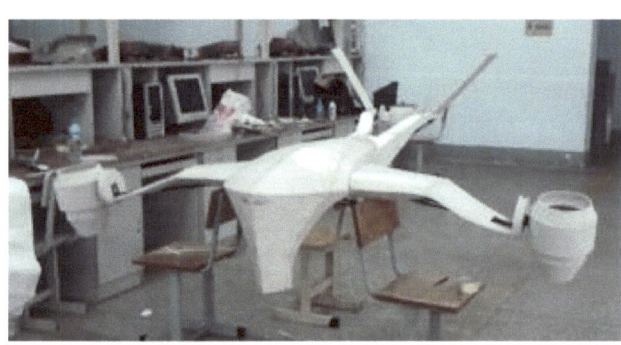

     Looks familiar.

Nanjing CK-20 High-Speed Target Drone

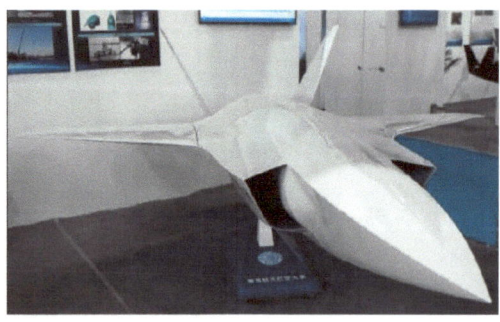

NAV Jet-powered Stealth UAV

NAV has designed two stealth UAVs that share the same configuration, and only differ from each other in propulsion. Made of composite material, the basic design adopts canard configuration and has radar cross section less than 0.01 m2. The two-blade propeller-driven version is air-launched by large transport aircraft or bombers mainly as a drone, and thus does not have landing gear. The two-stroke gasoline-powered engine is a pusher configuration and the engine is equipped with a FADEC system, which also equips the other jet-powered version for other applications such as surveillance and reconnaissance.

NAV WF170 Tomahawk Simulation Drone

This drone is designed to simulate a Tomahawk cruise missile to train air defense forces. The only one major visual difference is the inlet of WF170 is located above the fuselage, while that of the real missile is located below the fuselage. The speed of WF170 is 180 m/s.

NJUAV LY-Z270

LY-Z270 is a fixed-wing UAV developed by NJUAV in twin-boom layout with high-wing configuration and twin tails. Propulsion is provided by a two-blade propeller driven by a tractor engine mounted in the nose, and landing gear consists of a pair of skids.

NUAA "Cloud Leopard"

Clouded Leopard (Yun-Bao, Yunbao, 云豹) is a very little-known UAV. First revealed in 2007, it utilizes GPS/RA semi-autonomous landing technology, so it can land conventionally via taxiing during recovery instead of using parachute.

NUAA FX500 "Sky Saker" Rapid Artillery-Directing UAV

Designed for rapid artillery directing, especially for the long range MRLS (~300km). It features short straight wings and a V tail. It is powered by a small turbojet with the S-shaped air intake located above the rear fuselage. It can have the options of SAR and SATCOM installed in its head and nose, datalink antenna on its back, as well as EO turret underneath its head. Transported and launched from the launch vehicle using RATO.

Takeoff weight – 500 kg
Speed - 800 km/hr
Altitude - 9,000 m
Payload - 50 kg
Endurance - 2 hr

NUAA FY-E Reconnaissance UAV

FY-E UAV is a jet powered high-speed reconnaissance UAV. FY-E adopts some stealth features and can be deployed as target designator for precision guided munitions with semi-active laser terminal guidance. Externally, FY-E visually resembles AGM-86 ALCM, except FY-E adopts V-tail. The inlet of FY-E also has a different shape.

NUAA "Long Arrow" UAV

Long Arrow (Chang-Jian, 长箭, often abbreviated as CJ) is a fixed wing, jet powered UAV. Computer aided design was widely used. The engine inlet was designed via CAD, and constructed with fiber glass plastic that reduced radar cross section and improved stealth.

NUAA RKL 165 UAV

RKL 165 UAV is a UAV specifically designed to generate false targets to jam airborne early warning systems. RKL 165 is part of the QD550 electronic warfare system (EWS) under Project 995 (995工程). QD550 EWS of Project 995 is an integrated EWS covering large sector, specifically developed to counter E-2T, which consists of four subsystems, RCL 302, RCT 321, RCZ 301 and RKL 165. Technologies developed for RKL 165 have been successfully adopted by other UAVs developed by NUAA, such as Clouded Leopard UAV, BZK-002 UAV and Long Arrow UAVs.

NUAA WZ-1 Soar Bird UAV

Similar to Northrop Grumman Fire Scout. Capable of autonomous flight, various surveillance sensors, GPS and laser altimeter.

Operation radius - 150 km
Altitude - 3,000 m
Max speed - 125 km/h
Weight - 900 kg
Payload - 30 kg
Endurance - 4 hr

NUST "Sharp Sword" Carrier-based UAV

Experimental UAV for two purposes: gaining experience for aircraft carrier operations, and for UCAV. Sharp Sword is a miniature version of Sukhoi Su-35, which the UAV aerodynamics is based on, with an addition of tailhook. Sharp Sword (Li-Jian or Lijian, 利剑, not to be confused with another Chinese UAV with the same name, AVIC 601-S Sharp Sword).

Oxai B4 UAV

Powered by lithium batteries. B4 UAV utilizes Canadian autopilots and made it public debut at International UAV competition in Beijing. B4 UAV is a flying wing design with winglets and has tricycle landing gear system. Propulsion is provided by a propeller driven by a pusher engine.

   Cruise speed - 120 km/hr
   Altitude - 6 km
   Wingspan - 6 m
   Payload - 7 kg
   Endurance - 4 hr

SG-1 "Star Glory" UCAV –

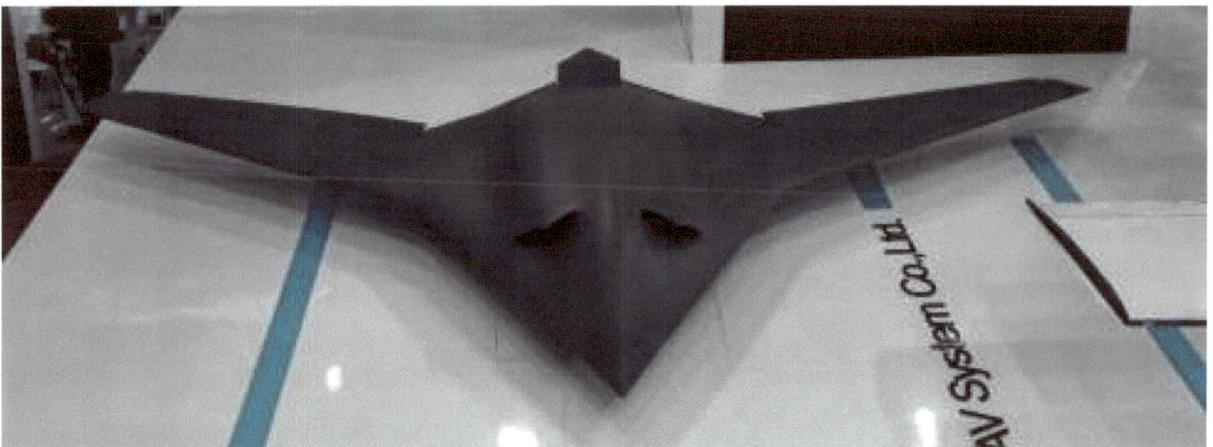

Built by Star Systems with two engines.

Speed - 650 km/h;
Altitude - 12,000 m
Range - 2,000 km
Take-off weight - 4,000 kg
Payload - 400 kg
Length - 6.8 m
Wingspan - 15 m
Endurance - 8 hours

SG-1 "Star Shadow" UCAV –

Built by Star Systems with two engines. Improved version of "**Star Glory**".

Altitude - 15,000 m
Take-off weight - 4,500 kg
Length - 5 m
Wingspan - 7.3 m
Endurance - 12 hours

SULA89 Kamikaze Drone

Derived from there was an older variant known as "Ferocious Tiger." The drone is fired eight pipes of large diameter on the roof of a Yanjing YJ2080C all-terrain vehicle. A warhead of more than 2 kg crashes into a target at a speed of 180 km/h. The vehicle has 12 pneumatic launch tubes and the other 4 tubes carry smaller **SULA30** reconnaissance drones.

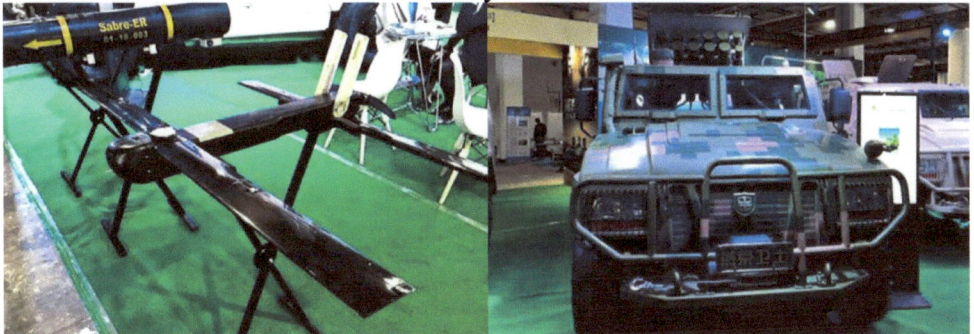

SYAC "Divine Eagle" Counter Stealth HALE UCAV

AKA **Shen-Diao**, 神雕. Twin boom, twin tail, largest UAV as of 2015. Has data link to pass targeting data to a VLRAAM-armed fighter (very-long-range air-to-air missile).

Speed - Mach 0.8
Altitude - 25 km
Wingspan - 40 to 50 m

SYAC "Spider-Man" ZZX Hunter Drone

Disables another drone by firing 16-m^2 (172 sq ft) web. Also performs surveillance and reconnaissance.

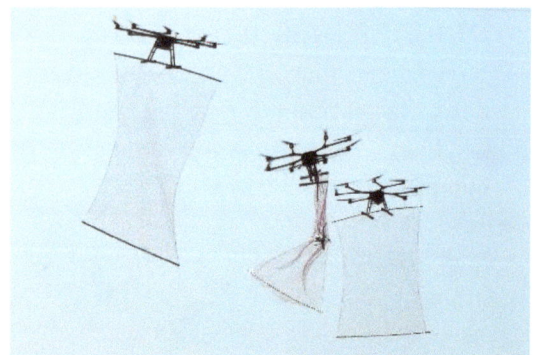

SYAC XLB "Patroller" Fixed-Wing UCAV –

AKA XLB Xun-Luo-Bing or XLB 巡逻兵. High wing config, tricycle landing gear and T-tail. Two-blade prop mounted in the nose. Powered by fuel cell.

Speed – 150 km/h
Altitude - 5,000 m
Weight – 240 kg
Payload – 30 kg
Endurance - 3-5 hours

Tengoen TB001 "Twin-Tailed Scorpion" UCAV

Twin-boom, twin-engine airframe. Can carry 220 pounds of ordnance on 8 hardpoints.

Range - 3,700 miles
Length - 33 m
Wingspan - 66 m
Endurance - 35 hr

Weapon
Blue Arrow missiles

V750 UCAV

Designed to operate autonomously and follow a preplanned flight plan as well as by the ground control station from 150 km. Can carry 2 missiles.
Speed - 161 km/h
Takeoff weight - 757 kg
Payload - 80 kg
Endurance - 4 hr

UVS U650-A2 Large Amphibious UAV

Amphibious with retractable landing gear, four hardpoints, water takeoff landing run 500 m.
Range - 2,000 km
Speed - 180 km/h
Altitude - 7,000 m
Length - 5.85 m
Wingspan - 12.4 m
Endurance - 12 hour
VVP Soft VVP-9066

WBZY BW-I Target Drone

BW-I is a drone designed to simulate low altitude subsonic cruise missile. BW-I has a total of eleven control surfaces with a pair of the smallest installed near the nose section of the fuselage. The midsection of the fuselage of BW-II has a total of five control surfaces installed, with three larger trapezoid shaped control surfaces in a reverse T configuration, and in the belly of the midsection, there are two smaller rectangular control surfaces. At the tail section of the fuselage, there are four rectangular control surfaces in cruciform and these control surfaces are also the largest.

   Speed - 0.7 – 0.9 Mach
   Altitude - 100 – 300 m
   Range - > 25 km
   Course accuracy - better than 15°
   Height accuracy - better than 50 m
   Engine - solid rocket motor

Winhye UCAV

Winhye has developed a UCAV that is derived from a reconnaissance UAV, and hence the complete name is Reconnaissance Strike Integrated UAV (Cha Da Yi-Ti Wu-Ren-Ji, 察打一体无人机). There are actually two versions of this UCAV, and both look similar to Boeing *Insitu ScanEagle* due to the identical layout the three UAV share. What differs the two versions of Winhye UCAV from each other is the winglets. For one version, the winglets point upward, while for the other, its winglets point downward.

> Range - 550 km
> Speed - 90 – 110 km/hr
> Altitude - 5 km
> Length - 2.65 m
> Wingspan - 2.2 m
> Payload - 15 L
> Max takeoff weight - 20 – 25 kg
> Endurance - 5 hr
> Launch - catapult or taxiing
> Recovery - parachute or taxiing
> Power plant - 50 – 110 cc gasoline engine

WZ-7 "Soar Dragon" HALE Reconnaissance UCAV

AKA Chengdu **Xianglong** ("**Soaring Dragon**"), AKA "**Sky Wing**", AKA "**mini-Global Hawk**" has unique tandem joined-wing design gives the UAV a diamond-shaped appearance. Has sensors for targeting ships and cruise missiles. The EW version of the Soar Dragon potentially could operate alongside other Chinese Electronic Warfare aircraft, such as the Y-9 heavy transport jammer, the J-16D strike fighter, and JH-7 attack aircraft. Together, these aircraft could be used to creating a jamming field that would block or scramble enemy communications, radar systems and more.

Range - 7,000 km
Speed - 750 km/h
Altitude – 17.3 - 18 km
Takeoff weight - 10-20 t
Endurance - 10 hours
Length - 13 m
Wingspan - 20 m

WZ-2000 Reconnaissance/Strike UCAV

The Guizhou WZ-2000, also known as the WuZhen-2000 or WZ-9. Onboard remote sensors include a thermal imaging camera and synthetic aperture radar (SAR). Imagery is transmitted via satellite communications antenna mounted in the nose bulge. The WZ-2000B is an improved production variant.

Range - 2,400 km (1,500 mi, 1,300 nmi)
Combat range - 800 km (500 mi, 430 nmi)
Service ceiling - 18,000 m (59,000 ft)
Maximum speed: 800 km/h (500 mph, 430 kn)
Payload - 80 kg (176 lbs.)
Length - 7.5 m (24 ft 7 in)
Wingspan - 9.8 m (32 ft 2 in)
Max takeoff weight - 1,700 kg (3,748 lbs.)
Endurance - 3 hours
Armament -
   (2) KD2 air-to-surface missiles
   Additional number of ZD1 precision-guided bombs
Powerplant - (1) WS-11 turbofan, 16.9 kN (3,800 lbf) thrust (essentially the Russian Ivchenko AI-25 series)

Xian CCKW LJ-I Stealthy Target Drone

Xinying "Clairvoyance V" UAV

Clairvoyance V is a very little-known UAV developed by Xinying and was once briefly publicized at the developer's website, but the web page has since been removed. It's possible that Clairvoyance V has been adopted by Chinese military, which often resulted in the removal of information about the UAV from public view. This assumption is further strengthened by the fact that the developer's own work published reveals that Clairvoyance 5 incorporates **frequency-hopping spread spectrum**, a measure generally reserved for military, but usually not for civilian use, and all previous Clairvoyance series lacked this feature. Furthermore, the developer dubbed the UAV as an aerial reconnaissance platform

XYAST XYB-180 Reconnaissance UAV

XYB-180 UAV is a jet-powered UAV developed by XYAST. Externally, XYB-180 resembles a miniature jet fighter with conventional layout. The inlet is located above the fuselage at the root of the single vertical rudder, and leading-edge extension is also incorporated. XYB-180 can be either vehicle launched or rocket assisted launched.

Max speed - 650 km/hr
Cruise speed - 360 – 480 km/hr
Weight - 56 kg
Endurance - 1 hr

XYAST KGXY-180

KGXY-180 is a little-known Chinese stealth UAV developed by XYAST and has since entered service with Chinese military. KGXY-180 is developed from XYB-180 and like XYB-180, it is jointly developed by XYAST and Wei-Nan Normal University(渭南师范学院) (WNNU) The existence of KGXY-180 is reveal when WNNU was won the first place of Military Science and Technology Advancement Award in 2011 for its successful development of KGXY-180 stealth UAV.

XY Aviation UR-J1-001 "Blue Arrow" BA-270

UR-J1-001 Blue Arrow (Lan-Jian or Lanjian, 蓝箭) BA-270 is a fixed-wing HALE (high-altitude long-endurance) UAV developed by XY Aviation in conventional layout with V-tail and tricycle landing gear. Propulsion is provided by a two-blade propeller driven by a pusher engine. UR-J1-001 Blue Arrow BA-270 is somewhat smaller than General Atomics MQ-9 Reaper and has very similar layout to that of MQ-9, but there is no bulge at the nose of the fuselage of UR-J1-001 Blue Arrow BA-270 to house the large satellite communication antenna on MQ-9, but on the wing tips of UR-J1-001 Blue Arrow BA-270, there are winglets that are absent on MQ-9.

- Wingspan - 7.188 m
- Wing area - 3 m2
- Length - 3.795 m
- Height - 1.202 m
- Empty weight - 100 kg
- Fuel - 72 kg
- Payload - 30 kg
- Max take-off weight - 202 kg
- Max speed - 180 km/hr
- Cruise speed - 130 km/hr
- Normal operating height - 4 km
- Endurance - 12 hr
- Max wind scale allowed for operation - 6
- Power plant - Limbach Flugmotoren L253E piston
- Launch & recovery - taxi

ZHZ TD220 Coaxial Unmanned Helicopter -

Range – 150 km
Speed – 100 km/h
Altitude – 3.5 km
Takeoff weight – 350 kg
Payload – 50 kg

ZJ-100 UAV

ZJ-100 has a cylindrical fuselage with a conical nose. There are four control surfaces mounted at the empennage. The UAV has been referred to by many Chinese internet sources as ZJ-100 long range missile, but the actual name on the UAV clearly states that it is a UAV instead, though it can certainly be converted to a missile by installing a warhead. The official name on ZJ-100 UAV at the airshow is ZJ-100 High Speed Rocket Drone. The UAV presumably is used to simulate missiles for air defense systems, though it can also be used as other missions such as reconnaissance when a different payload is carried.

www.ingramcontent.com/pod-product-compliance
Lightning Source LLC
Chambersburg PA
CBHW051210220526
45473CB00003B/973